STUDY GUIDE AND SOLUTIONS MANUAL FOR STUDENTS

Prepared by David Freifelder,
University of California, San Diego

To Accompany

General GENETICS

LEON A. SNYDER   DAVID FREIFELDER   DANIEL L. HARTL

Jones and Bartlett Publishers, Inc.
BOSTON   PORTOLA VALLEY

Copyright © 1985 by Jones and Bartlett Publishers, Inc. All rights reserved. No part of the material protected by this copyright notice may be reproduced or utilized in any form, electronic or mechanical, including photocopying, recording, or by any information storage or retrieval system, without written permission from the copyright owner.

Editorial offices: Jones and Bartlett Publishers, Inc., 30 Granada Court, Portola Valley, CA 94025.

Sales and customer service offices: Jones and Bartlett Publishers, Inc., 20 Park Plaza, Boston, MA 02116.

ISBN 0-86720-054-5

Printed in the United States of America

Printing number (last digit): 10 9 8 7 6 5 4 3

INTRODUCTION

Genetics is usually difficult for the beginner. Notions of probability are unfamiliar, and the terminology may seem intimidating. However, genetics is an inherently logical science, and once the logic is appreciated, things become considerably clearer. Certain topics are recognized by all teachers of genetics as being the more difficult ones, and these topics are often reviewed and expanded on in the classroom. In writing *General Genetics*, the authors attempted to keep the book to a length that could be completed in a one-semester course. With this goal, it was not possible to offer extensive reviews of basic biology and to give repeated examples of particular topics. Recognition of the value of such additional material provided the impetus for my writing the *Study Guide and Solutions Manual for Students*. The Guide is divided into 16 chapters, each one corresponding to the chapter with the same number in *General Genetics*. Each chapter is subdivided into eight parts: Chapter Summary, Bold Terms, Additional Information, Drill Questions, Answers to Drill Questions, Additional Problems, Answers to Additional Problems, Solutions to Problems in the Text. The student should use these parts in the following way:

Chapter Summary. This section summarizes the major concepts in the chapter, which are presented usually in the same order as in the text. The student should simply scan this section to be sure that each point is remembered.

Bold Terms. In the text, important terms and definitions are printed in boldface letters when they are first defined. These terms represent the vocabulary of genetics and *must* be understood. This section of the Guide lists all of the terms. The student should not proceed with further study unless each term can be defined and understood. Definitions can be found in the Glossary of *General Genetics*, and the Index to the text can be used to find the first use of the term.

Additional Information. Most students have difficulty with certain features of genetics and may need additional help. In this section tables, hints, examples of solving problems, and short reviews of biological concepts and genetic principles are provided, so that the student having difficulty can find help out of the classroom. Furthermore, the student interested in extending his or her knowledge beyond what is given in the text will find in this section some material that is either slightly more advanced or simply something that was omitted in the interest of keeping the text fairly short.

Drill Questions. These are simple questions that test the most elementary and basic facts and concepts. The student must be sure that each question can be answered with confidence. If this is not the case, he or she should return to the text for review. Proceeding to subsequent chapters and attempting to work the problems at the end of the same chapter without such a review will surely be a waste of time if all of these questions cannot be answered.

Answers to Drill Questions. Answers, with explanations when appropriate, are given in this section.

Additional Problems. This small collection contains problems of two types. Some merely reexamine a few of the more important concepts illustrated by the problems in the text. Others are a little more challenging and may go beyond what is expected of the typical beginning student of genetics.

Answers to Additional Problems. Complete answers to the Additional Problems are given in this section.

Solutions to Problems in Text. Answers to all problems in *General Genetics* are given in the back matter at the end of the book. However, to save space, the answers are quite abbreviated, often a single word or number. The final section of each chapter of this Guide contains detailed solutions to each of the questions in the text. These solutions should enable the student to fully comprehend the principles examined in the problems. This section may be consulted if a student is completely unable to understand a problem, but a fair chance should be given to each problem before looking.

January, 1985　　　　　　　　　　　　　　　　　　　　　　　David Freifelder
San Diego, California

CONTENTS

CHAPTER 1
The Elements of Heredity and Variation ... 1

CHAPTER 2
Genes and Chromosomes ... 17

CHAPTER 3
Gene Linkage and Chromosome Mapping ... 33

CHAPTER 4
The Chemical Nature and Replication of the Genetic Material ... 55

CHAPTER 5
The Molecular Organization of Chromosomes ... 67

CHAPTER 6
Variation in Chromosome Number and Structure ... 81

CHAPTER 7
Genetics of Bacteria and Viruses ... 95

CHAPTER 8
Mechanisms of Genetic Exchange ... 111

CHAPTER 9
Gene Expression ... 123

CHAPTER 10
Mutation and Mutagenesis ... 137

CHAPTER 11
Regulation of Gene Activity ... 153

CHAPTER 12
Genetic Engineering ... 171

CHAPTER 13
Somatic Cell Genetics and Immunogenetics ... 181

CHAPTER 14
Population Genetics ... 195

CHAPTER 15
Genetics and Evolution ... 209

CHAPTER 16
Quantitative Genetics ... 221

CHAPTER 1

The Elements of Heredity and Variation

CHAPTER SUMMARY

Inherited traits are determined by particulate elements called genes, as first recognized by Mendel. A gene has different forms, called alleles. In a higher plant or animal the genes are present in pairs, one member of each pair having been transmitted from the maternal parent and the other member from the paternal parent. The specific allelic or genetic composition of an individual is called its genotype; the observable properties that result from interaction of the genotype with the environment is termed its phenotype. If the two alleles of a pair are the same (for example, *AA* or *aa*), the organism is said to be homozygous with respect to that gene; if the alleles are different (*Aa*), the organism is heterozygous. It is common that the phenotype of a heterozygote is the same as that of one of the homozygous combinations. When this is the case, the member of the pair also present in the homozygote is said to be dominant over the other, and the nonexpressed allele is termed recessive.

The principles of inheritance were inferred by Mendel from the study of pairs of contrasting traits in pea plants. The organisms produced by a mating constitute the F_1, or first filial generation. If these individuals mate, the progeny constitute the F_2 or second filial generation. In crosses of the type *AA* x *aa* in which only one gene was considered (monohybrid crosses), Mendel observed that in the F_2 generation obtained by self-pollination of the F_1 hybrids, the phenotypes occurred in the ratio 3 dominant:1 recessive. From further crosses it was determined that a genotypic ratio of 1 dominant homozygote:2 heterozygotes:1 recessive homozygote underlies all 3:1 phenotypic ratios. The principle inferred from these results is that in the processes leading to formation of gametes (reproductive cells) the members of an allelic pair are segregated randomly into different gametes, and the gametes participate in fertilization also in random combinations. In crosses of the type *AABB* x *aabb* (dihybrid crosses) Mendel found phenotypic ratios of 9:3:3:1 in the F_2 generation and recognized that they represent the random combination of populations whose phenotypic ratios are both 3:1. From these results his second principle was inferred, namely, that in the transmission of two or more pairs of alleles the members of different pairs are assorted into the gametes independently of each other, with the result that each combines at random during fertilization. The random processes occurring in the formation of gametes and their union at fertilization follow the simple rules of probability, which provide the basis for predicting outcomes of genetic crosses. In some organisms, for example, humans, it is not possible to perform crosses. In such cases genetic analysis is carried out by study of a diagram of the phenotypes of members of a family through several generations. Such a diagram is called a pedigree, and pedigree analysis refers to the determination of the genotypes of the family members and the probability of a particular genotype being associated with a particular member.

Exceptions to the dominance of one member of a pair of alleles over the other occur frequently in most organisms. For example, intermediate phenotypes occasionally arise because less of a gene product (which causes the phenotype) is made when only one wildtype allele is present than when two copies are present. This phenomenon is called partial dominance. An additional exception to simple dominance occurs when the two different alleles in a heterozygote are both fully expressed, a phenomenon called codominance. Other types of exceptions to the pattern of inheritance expected from Mendel's principles of segregation and independent assortment are known. For example, one gene may affect the expression of other genes; this phenomenon, which is called epistasis, is fairly common and complicates genetic analysis.

BOLD TERMS

agglutination, alkaptonuria, alleles, antibody, antigen, antiserum, backcross, carrier, codominance, dihybrid, dominant, epistasis, first filial generation, F_1, F_2, gamete, gene, genotype, heredity, heterozygous, homozygous, hybrid, inborn errors of metabolism, monohybrid, natural selection, pedigree, phenotype, principle of independent assortment, principle of segregation, Punnett square, recessive, reciprocal cross, second filial generation, segregation, siblings, sibs, testcross, true-breeding, variable expressivity, variable penetrance, zygotes.

ADDITIONAL INFORMATION

Table 1 summarizes terms whose meanings and uses must be understood before proceeding to more advanced material. Here is an example of the use of these terms and symbols: A heterozygous animal, whose genotype is *Bb*, has the black hair phenotype because the allele *B* is dominant and allele *b* is recessive. The homozygous dominant animal (*BB*) has black hair; the homozygous recessive animal (*bb*) has white hair.

Note that the notation *AaB-* indicates that the genotype can be *AaBB* or *AaBb*. Similarly, *A-B-* represents four genotypes: *AABb*, *AABB*, *AaBB*, and *AaBb*.

Writing down the genotypes for a cross with one or two genes segregating can be easily done with a checkerboard or Punnett square with the gametes written along the two axes of the square. For example, with one gene the genotypes produced in the cross *Aa* x *Aa* are those shown in Figure 1.

Table 1 Some genetic terms applied to hair color of an animal

Term	Definition	Examples
Allele	Different forms of a gene	B or b
Phenotype	An observed property	Black hair or white hair
Genotype	Genetic constitution of an organism	BB, Bb, or bb
Heterozygous	Refers to the genotype in which the alleles differ	Bb
Homozygous	Refers to a genotype in which the alleles are the same	BB or bb
Dominant allele	An allele expressed in a heterozygote	B
Recessive allele	An allele expressed only in a homozygote	b

Figure 1

	A	a
A	AA	Aa
a	Aa	aa

With two genes segregating (for example, AaBb x AaBb), there are 16 squares in the box, as shown in Figure 2.

Figure 2

	AB	Ab	aB	ab
AB	AABB	AABb	AaBB	AaBb
Ab	AAbb	AAbb	AaBb	Aabb
aB	AaBB	AaBb	aaBB	aaBb
ab	AaBb	Aabb	aaBb	aabb

If you try to form a Punnett square with three or more genes, you will discover that it is exceedingly cumbersome. In this case a branching diagram (sometimes called a tree diagram) is the method of choice. Table 2 shows how the gametes are enumerated and Table 3 shows a diagram for determining the genotypes. In Table 3, note that one is dealing with a diploid state instead of haploid gametes, so the diagram is set down by starting with all possible diploid conditions for the first gene pair (Step 1). We continue in the same way for as many branches as there are gene pairs. In this table there are two more gene pairs, whose combinations are worked out in Steps 2 and 3. The genotype is finally summarized for each path in Step 4.

Table 2 Branch diagram for formation of gametes from an AaBbCc parent

---

Step 1   Step 2   Step 3   Step 4

```
                    ┌── C ──── ABC
              ┌ B ──┤
              │     └── c ──── ABc
         A ───┤ or
              │     ┌── C ──── AbC
              └ b ──┤
                    └── c ──── Abc
  or
                    ┌── C ──── aBC
              ┌ B ──┤
              │     └── c ──── aBc
         a ───┤ or
              │     ┌── C ──── abC
              └ b ──┤
                    └── c ──── abc
```

A persistent problem in all genetic analysis is being sure that all gametes have been written down. If this listing is done in orderly fashion, usually none will be omitted. Nonetheless, it is easy to make a mistake, so it is valuable to have a check on the number of gametes. Review Drill Question 4 to be sure that you know how to check whether the number you have listed is correct.

The effects of codominance and epistasis often cause confusion. An important point is that codominance increases the number of phenotypic classes, whereas epistasis decreases the number of classes. These effects can be seen most easily in Table 4. Here we consider a cross $TtRr \times TtRr$ in four cases: T dominant to t and R dominant to r; T dominant to t and R codominant with r; both T and R codominant with their alleles; tt epistatic to

Table 3 Twenty-seven different F_2 genotypes

```
Step 1          Step 2          Step 3          Step 4

                                  CC ─────── AABBCC
                BB ─────────── Cc ─────── AABBCc
                                  cc ─────── AABBcc
                                  CC ─────── AABbCC
AA ─────────── Bb ─────────── Cc ─────── AABbCc
                                  cc ─────── AABbcc
                                  CC ─────── AAbbCC
                bb ─────────── Cc ─────── AAbbCc
                                  cc ─────── AAbbcc
                                  CC ─────── AaBBCC
                BB ─────────── Cc ─────── AaBBCc
                                  cc ─────── AaBBcc
                                  CC ─────── AaBbCC
Aa ─────────── Bb ─────────── Cc ─────── AaBbCc
                                  cc ─────── AaBbcc
                                  CC ─────── AabbCC
                bb ─────────── Cc ─────── AabbCc
                                  cc ─────── Aabbcc
                                  CC ─────── aaBBCC
                BB ─────────── Cc ─────── aaBBCc
                                  cc ─────── aaBBcc
                                  CC ─────── aaBbCC
aa ─────────── Bb ─────────── Cc ─────── aaBbCc
                                  cc ─────── aaBbcc
                                  CC ─────── aabbCC
                bb ─────────── Cc ─────── aabbCc
                                  cc ─────── aabbcc
```

Table 4 Effect of dominance and epistasis on phenotypic classes from a dihybrid cross

Frequency of genotypes	T and R dominant	T dominant R codominant	T and R codominant	T and R dominant, tt epistatic to R locus
1/16 TTRR	⎫	⎫	1	⎫
2/16 TtRR	⎬ 9 T-R-	⎬ 3 T-RR	2	⎬
2/16 TTRr	⎪	⎭	2	⎪ 9 T-R-
4/16 TtRr	⎭	⎫ 6 T-Rr	4	⎭
1/16 TTrr	⎫ 3 T-rr	⎫ 3 T-rr	1	⎫ 3 T-rr
2/16 Ttrr	⎭	⎭	2	⎭
1/16 ttRR	⎫	1 ttRR	1	⎫
2/16 ttRr	⎬ 3 ttR-	2 ttRr	2	⎬ 3 ttR 1 ttrr
1/16 ttrr	1 ttrr	1 ttrr	1	⎭

R and r (recessive epistasis). The gametes produced by a $TtRr$ individual are TR, tR, Tr, and tr. The F_2 zygotic combinations and their relative frequencies can be obtained from a Punnett square. They are 1/16 $TTRR$, 2/16 $TtRR$, 2/16 $TTRr$, 4/16 $TtRr$, 1/16 $TTrr$, 2/16 $Ttrr$, 1/16 $ttRR$, 2/16 $ttRr$, and 1/16 $ttrr$.

DRILL QUESTIONS

1. What gametes can be formed by an individual whose genotype is Bb?

2. What gametes can be formed by two individuals whose genotypes are $AaBb$ and $AaBB$?

3. How many different gametes can be formed by an organism with genotype $AABbCcDdEe$?

4. How many different gametes can be formed by an organism that is heterozygous for m genes and homozygous for n genes?

5. If R is dominant to r, how many different phenotypes are present in the progeny of a cross between   it Rr and Rr? In what ratios?

6. If there is no dominance between R and r, how many different phenotypes result from the cross $Rr \times Rr$? In what ratios?

7. What can be said, with respect to homozygosity and heterozygosity, about the genotype of an individual that is true-breeding for a particular trait?

8. A true-breeding, long-tailed animal is crossed with a short-tailed animal that does not breed true. The progeny are all long-tailed. Which animal is heterozygous, which is homozygous, and which tail length is dominant?

9. A true-breeding, red-haired aardvark is crossed with a true-breeding, white aardvark. All of the F_1 are pink. What phenomenon does this probably represent?

10. A true-breeding, red-skinned worm mates with a blue-skinned worm. All progeny are purple. When purple worms are interbred, red-, purple-, and blue-skinned worms result, in the ratio 1:2:1. What phenomenon does this represent?

11. A coin is flipped and comes up heads. What is the probability that the next flip will yield heads? What about the third and fourth flips?

12. What is the probability of getting two heads in a row in two flips of a coin?

13. In a pedigree a homozygous recessive results from the mating of a heterozygote and a parent with the dominant phenotype. What does this tell you about the genotype of that parent?

14. In a pedigree two parents having a dominant phenotype produce nine offspring. Two have the recessive phenotype. What does this tell you about the genotype of the parents?

15. In a pedigree one parent has a dominant phenotype and the other has a recessive phenotype. Two offspring result and both have the dominant phenotype. What genotypes are possible for the other parent and with what probabilities?

16. Pedigree analysis tells you that a particular parent may have the genotype *AABB* or *AABb,* each with the same probability. What is the probability of that parent producing an *Ab* gamete? What is the probability of producing an *AB* gamete?

ANSWERS TO DRILL QUESTIONS

1. *B* and *b.*

2. *AaBb*: *AB   Ab   aB   ab*. *AaBB*: *AB   ab*.

3. Multiply the number of possibilities for each gene, that is, (1)(2)(2)(2)(2) = 16.

4. The homozygous genes are of no concern since they can produce only one type. Therefore, the number is 2^m.

5. Two phenotypes. The dominant and recessive phenotypes are in the ratio 3:1.

6. Three phenotypes corresponding to the three genotypes *RR, Rr,* and *rr,* in the ratio 1:2:1 .

7. The individual is homozygous for all genes that determine the trait.

8. The long-tailed and short-tailed animals are homozygous dominant and heterozygous respectively.

9. Partial dominance.

10. Codominance.

11. The probability of heads is always 1/2 for a single flip; the coin has no memory of what has happened.

12. This problem differs from problem 11 in that we require not only that the second flip gives heads but also that the first gives heads. Since the probability of each flip yielding heads is 1/2, the probability of two heads in a row is (1/2)(1/2) = (1/4).

13. The parent must carry one copy of the recessive allele. Since the parent has the dominant phenotype, the parent must be a heterozygote.

14. Both parents must be heterozygotes.

15. The other parent can be either homozygous dominant or heterozygous, each genotype with the same probability. The fact that no recessives result might make you suspect that the parent is a homozygous dominant, but a sample size of two offspring is not sufficient to rule out heterozygosity.

16. The probability of the parent being $AABb$ is 1/2, and if that is the parental genotype, the probability of producing a gamete Ab is 1/2. Therefore, the overall probability of producing an Ab gamete is (1/2)(1/2) = 1/4. Since only AB and Ab gametes can be formed, the probability of producing AB is 1 − 1/4 = 3/4. The latter can also be calculated in the following way. The probability of the $AABB$ genotype is 1/2 and the probability of producing AB is 1; therefore, the contribution to the overall probability of producing AB is 1(1/2). From the $AABb$ the contribution is (1/2)(1/2) = 1/4. Therefore, the total probability of producing AB is the sum of the two contributions, or 1/2 + 1/4 = 3/4.

ADDITIONAL PROBLEMS

1. A plant heterozygous for four independently assorting genes ($AaBbCcDd$) is self-fertilized. Determine the expected frequency of the following genotypes in the progeny of such a plant: (1) $aabbccdd$; (2) $aabbCcDd$; (3) $AaBbCcDd$; (4) $AABBCCdd$. Then, assuming incomplete dominance, how many different phenotypes will be present in the progeny?

2. Two camels known to differ by four gene pairs are mated. Assuming that the parents were homozygous for each of the four genes, that all four pairs segregate independently, and that dominance is complete, what is the probability that an F_2 offspring will be homozygous for all four recessive genes?

3. Two phenotypically normal parents have produced one albino child. What is the chance that their third child will be an albino?

4. The Chinese primrose has a center whose color differs from that of the petals. This center is normally yellow and medium sized. Variants include a large yellow center (Queen), white center (Alexandra), and blue center (Blue Moon). The results of certain crosses are the following:

Parent	F_1	F_2
Normal x Alexandra	Alexandra	3 Alexandra : 1 normal
Alexandra x Queen	Alexandra	Not reported
Blue Moon x Normal	Normal	3 Normal : 1 Blue Moon
Queen x Blue Moon	Blue Moon	Not reported

(a) Arrange the phenotypes in order of dominance.
(b) What F_2 is expected for crosses with unreported results?
(c) List all of the genotypes possible for color of the center.

5. In the cross $AaBb \times AaBb$, A and a are codominant and B is dominant to b. List each phenotype and its frequency in the population.

6. Mating of a brachyuric (short-tailed) mouse with a homozygous long-tailed mouse yields a 1:1 ratio of short- and long-tailed mice. Crossing two short-tailed mice gives a 2:1 (not 3:1) ratio of short-tail to long-tail. Explain these results and, in particular, state whether the gene causing short tail is dominant or recessive.

7. An ear of corn was obtained from the self-pollination of a plant heterozygous for a recessive allele w, resulting in albinism when homozygous. As expected, 1/4 of the resulting seedlings were albino and soon died, because they could not photosynthesize. The remaining green plants were allowed to interpollinate freely, and all of the seeds were saved. In the population of seedlings grown from these seeds the next year, what proportion would be expected to be albino?

8. The presence of at least one dominant allele of each of three genes A_1, C_1, and R is required for the production of color in kernels of maize; that is, the genotypes A_1-C_1-R- are colored and all others are colorless. A plant of a strain with colored kernels is crossed with three tester plants, with the following results: (1) In the cross with $A_1A_1c_1c_1rr$, 50 percent of the kernels were colored. (2) In the cross with $a_1a_1C_1C_1rr$, 25 percent of the kernels were colored. (3) In the cross with $a_1a_1c_1c_1RR$, 50 percent of the kernels were colored. What is the genotype of the plant being tested?

9. Thousands of different mutants have been isolated in Drosophila. One of these is the recessive vestigial wing mutation vg. Given two phenotypically normal flies, one female and one male, design a test to determine if either of them are carriers of the vg allele.

ANSWERS TO ADDITIONAL PROBLEMS

1. (1) $(1/4)(1/4)(1/4)(1/4) = 1/256$; (2) $(1/4)(1/4)(1/2)(1/2) = 1/64$; (3) $(1/2)(1/2)(1/2)(1/2) = 1/16$; (4) $(1/4)(1/4)(1/4)(1/4) = 1/256$; (3)(3)(3)(3) = 81.

2. Let us choose a hypothetical arrangement of genes (the exact array is unimportant), for example, $AAbbCCDD \times aaBBccdd$. The F_1 cross would then be $AaBbCcDd \times AaBbCcDd$. The probability of any gamete having either A

or a is 1/2. Similarly, the probability of any gene of a pair being in a particular gamete is 1/2. Since the genes segregate independently, the total probability of occurrence of several genes in a gamete is the product $(1/2)^4$. The probability of union of such gametes is the probability of joint occurrence of two independent events or $(1/2)^4(1/2)^4 = 1/256$. Alternatively, the probability that the F_2 offspring will be homozygous recessive at the A locus is 1/4 and at all four loci is $(1/4)^4 = 1/256$.

3. An albino individual must be homozygous (cc) for the albinism gene. Since in this case both parents are phenotypically normal, they must be heterozygous (Cc) for albinism. Such a monohybrid cross ($Cc \times Cc$) had a 1/4 chance that any given child will have the homozygous recessive cc genotype.

4. (a) The first cross shows that Alexandra is dominant to Normal. The second shows that Alexander is dominant to Queen. The third shows that Normal is dominant to Blue Moon. The fourth indicates that Blue Moon is dominant to Queen. That is, $A > N$, $A > Q$, $N > B$, and $B > Q$, or $A > N > B > Q$. (b) 3:1 for each. (c) Clearly, there are multiple alleles for this locus. The genotypes are: AA, AN, AB, AQ, NN, NB, NQ, BB, BQ, and QQ, or 10 in all.

5. Construct a Punnett square with the gametes AB, Ab, aB, and ab. BB and Bb will have the phenotype B, and bb will have the phenotype b. AA, Aa, and aa will have phenotypes A, Aa, and a, respectively. Therefore, the phenotypes and their frequencies are 3/16 AB, 6/16 AaB, 1/16 Ab, 2/16 Aab, 3/16 aB, and 1/16 ab.

6. Denote the normal by +/+ and the short-tailed mouse as $T/+$. The gametes are T and $+$ from the short-tailed individual and $+$ from the long-tailed individual. Thus, the F_1 consists of 1/2 $T/+$ and 1/2 +/+. Since two types appear in the F_1, the T gene must be dominant. If normal segregation were occurring, the second cross would yield 1/4 T/T, 1/2 $T/+$, and 1/4 ++. Assuming that T/T and $T/+$ have the same phenotype (which would be expected), the phenotypic ratio should be 3:1, not 2:1. The T/T individuals are clearly not present in the F_2. Therefore, the T/T combination must be a lethal. That is, the gene for a short tail is dominant in a heterozygote and lethal in a homozygote. Many mutations are known that behave in this way. They are called lethals.

7. The ratio of WW and Ww plants in the F_1 is 1:2. When freely intercrossed, WW plants will pollinate both WW and Ww, and Ww plants will pollinate both WW and Ww plants. Only a mating between Ww and Ww can produce a ww seed. Note that 2/3 of the F_1 are Ww, so 4/9 of the matings will be $Ww \times Ww$. Of

these matings, 1/4 of the seeds will be ww. Therefore, the fraction of the total number of seeds that is ww is $(4/9)(1/4) = 1/9$.

8. Consider cross 1. Inasmuch as the tester strain is AA, we cannot obtain any information about the allelic constitution of this gene in the plant being tested. However, the cross yields 50 percent colored kernels, which means that the tested plant had to be heterozygous for one of the C_1 and R genes and homozygous dominant for the other; in other words, the genotype must be either C_1c_1RR or C_1C_1Rr. If the plant had been heterozygous for both of these genes, the cross would instead have yielded 25 colored kernels. That was the result obtained from the second cross (the one with the $a_1a_1C_1C_1rr$ tester strain), which means that the tested plant must have been heterozygous for the R gene and also heterozygous for the A_1 gene. Therefore, the genotype of the plant must be $A_1a_1C_1C_1Rr$. This agrees with the result of cross 3 (the one with the $a_1a_1c_1c_1RR$ tester strain), from which 50 percent of the kernels were colored.

9. Perform a simple cross. If both flies carry the vg allele, a 3:1 ratio of normal to vestigial wings will result. If only one fly carries the vg allele, all flies in the F_1 will be normal. This would be true also if neither fly carried the allele. However, in the F_2, if one fly carries the vg allele, 1/16 of the flies in the F_2 will have vestigial wings.

SOLUTIONS TO PROBLEMS IN TEXT

1. (a) The parents are true-breeding and hence are homozygous for each gene. The genotypes of the parents are $DDHHww$ and $ddhhWW$. In a cross between these parents the F_1 plants will all have the genotype $DdHhWw$. To determine the types of gametes, prepare the branch diagram shown in below.

Text Problem 1

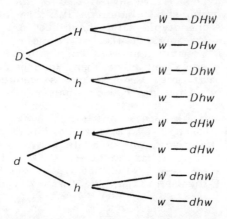

All gametes will be produced in equal proportions. (b) To answer this question, prepare a Punnett square. The genotype of the tall hairy white-flowered parent (*DDHHww*) can be produced only by the combination of gametes *DHw* and *DHw*. There are 64 combinations, so the proportion is 1/64. The phenotype of that parent can be produced by genotypes *DDHHww* (found in one box), *DDHhww* (found in two boxes), *DdHHww* (found in two boxes), and *DdHhww* (found in four boxes), so the proportion is 9/64. The genotype of the other parent can only be produced in one way, by combining *dhW* and *dhW*. The phenotype of this parent corresponds to two genotypes, *ddhhWw* found in two boxes) and *ddhhWW* (found in one box), so the proportion is 3/64.

2. (a) This can be obtained from a Punnett square. However, the number can be obtained by the following reasoning. Genes *A* and *B* will yield four phenotypes (those occurring in the 9:3:3:1 ratio). Because the *R-r* pair shows no dominance, there are three phenotypes associated with these alleles, namely, *RR, Rr,* and *rr*. Therefore, the number of phenotypic classes is 4 × 3 = 12. (b) The probability of *aabbRR* is (1/4)(1/4)(1/4) = 1/64. (c) Be sure you understand this problem. Homozygous for all three genes includes combinations such as *AABBrr, aabbRR, AAbbrr,* and so forth, of which there are eight. Since the probability of homozygosity of a single allele is 1/4, the proportion requested is the probability of each homozygous combination, which is (1/4)(1/4)(1/4), multiplied by 8 (the number of combinations, or 8/64 = 1/8.

3. (a) The dwarf parents must be heterozygotes to produce two types of offspring. If the condition were a recessive, they would be normal but carriers. Since they show the dwarf phenotype, achondroplasia must be dominant. (b) The order of birth is unimportant. Since both parents are heterozygous, the probability of a normal (homozygous recessive) child is 1/4.

4. Since a deaf (homozygous recessive) daughter is produced, both parents must be heterozygous (*Dd*). Therefore, each produces gametes *D* and *d* with equal probability and progeny *DD, Dd,* and *dd* in the ratio 1:2:1. The son is not deaf and hence cannot be *dd*. The ratio of genotypes with the normal phenotype is 1 *DD*: 2 *Dd,* so the probability of being heterozygous is 2/3.

5. (a) Assuming that the father is heterozygous, half of his gametes will contain the *D* allele, so the probability is 1/2 that the son will develop the disorder. (b) The son is less than 45, so we do not know whether he is a carrier. The probability is 1/2 that he is (as in part (a)), and, if so, half of his gametes will carry the *D* allele. Therefore, the probability that his son has the allele is (1/2)(1/2) = 1/4.

6. Since her paternal grandfather had phenylketonuria, her father has a probability of 1 of being heterozygous. Since her mother is presumably normal, the probability is 1/2 that she is a carrier and 1/4 that she will produce a p gamete. Since her husband's brother is phenylketonuric, both of his parents must be heterozygous. All members of his generation will be either PP, Pp, or pp in a 1:2:1 ratio. He is not pp, so the probability that he is a carrier is 2/3. Thus, the chance that he will produce a p gamete is $(1/2)(2/3) = 1/3$, and the probability of a pp child is the product of the mother producing a p gamete (1/4) and of the father producing a p gamete (1/3), or $(1/4)(1/3) = 1/12$.

7. Since each had an albino parent (cc) and assuming that the other parent was homozygous dominant (since albinism is rare), each is heterozygous (Cc). Therefore, the probability of an albino child is the probability of a homozygous recessive, which is 1/4. To produce two albino children, the probability is $(1/4)(1/4) = 1/16$. Since both homozygous dominants and heterozygotes are not albino, and the nonalbino:albino phenotypic ratio is 3:1, the probability of a nonalbino is 3/4. The probability of two nonalbino children is $(3/4)(3/4) = 9/16$. Thus, the probability of at least one child being albino is 1 minus the probability that none are albino, or $1 - 9/16 = 7/16$.

8. (a) Since both parents are true-breeding, their genotypes must be $RRBB$ and $rrbb$. Therefore, all F_1 progeny are $RrBb$, which are red. (b) The gametes are RB, Rb, rB, and rb. Construct a Punnett square, which will have 16 entries; 9 will have both an R and a B and be red, 6 will contain either an R or a B (but not both and will be brown, and one will contain neither R nor B and will be white.

9. The genotypes of the Leghorn and the Wyandotte are $CCII$ and $ccii$, respectively, so the genotype of the F_1 is $CcIi$. You may construct a Punnett square or use the following reasoning. To be colored, the ii combination and either CC or Cc must be present. The probability of ii is 1/4 and that of CC or Cc is 3/4, so the combined probability is $(1/4)(3/4) = 3/16$.

10. (a) To determine the proportion having the white phenotype, one need only consider the C and c alleles. Thus, the proportion is just that of a homozygous recessive, or 1/4. (b) Since each allele will appear in gametes with a probability of 1/2, the probability of having the F_1 genotype ($AaBbCc$) is $(1/2)(1/2)(1/2) = 1/8$. Since the F_1 has the dominant phenotype for each gene, which has a probability of 3/4 for each gene, the probability of the F_1 phenotype is $(3/4)(3/4)(3/4) = 27/64$. (c) The probability of the genotype $aabbcc$ is the product of the probabilities for homozygosity of each gene, or $(1/4)(1/4)(1/4) = 1/64$. For the genotype $aaBbcc$ the probability is $(1/4)(2/4)(1/4) = 1/32$.

11. (a) The genotype of the F_1 is *Aa Cc Rr Prpr*, which has a purple phenotype. (b) Since color requires just one dominant allele (both homozygous dominant or heterozygous are acceptable), which arises at a probability of 3/4, and red color requires a homozygous *pr pr* combination, which arises at a probability of 1/4, the probability of red in the F_2 is (3/4)(3/4)(3/4)(1/4) = 27/256. The fraction that will be purple is (3/4)(3/4)(3/4)(3/4) = 81/256. Therefore, the total fraction with color is (27 + 81)/256 = 108/256, and the fraction without color is 1 − 108/256 = 148/256 = 37/64. (c) In a testcross each allele from a heterozygous locus is produced in the progeny with a probability of 1/2. To get red, (1/2)(1/2)(1/2)(1/2) = 1/16. Purple will also be produced with a probability of 1/16, so colorless seeds will arise with a probability of 1 − 2(1/16) = 7/8.

12. In the F_2 the probabilities of colored and colorless are 3/4 and 1/4, respectively. Assuming color can be produced, the probabilities of black and brown are 3/4 and 1/4, respectively. The probabilities of solid and spotted are 3/4 and 1/4, respectively. Therefore, solid black is (3/4)(3/4)(3/4) = 27/64, spotted black is (3/4)(3/4)(1/4) = 9/64, solid brown is (3/4)(3/4)(1/4) = 9/64, spotted brown is (3/4)(1/4)(1/4) = 3/64, and albino is 1 − (27 + 9 + 9 + 3)/64 = 16/64.

13. The parents in generation I must be heterozygous. The progeny 2 and 4 can be either *BB* or *Bb* and have a probability of 2/3 of being *Bb*. Therefore, they produce *b* gametes with a probability of 1/3, so the progeny III-1 and III-2 each have a probability of 1/3 of being carriers. They each produce *b* gametes at a probability of (1/2)(1/3) = 1/6, so the probability of a *bb* offspring is (1/6)(1/6) = 1/36.

14. The combinations *B-, E-, bb,* and *ee* arise with probabilities 3/4, 3/4, 1/4, and 1/4, respectively. Therefore, black is (3/4)(3/4) = 9/16, brown is (3/4)(1/4) = 3/16, and yellow is 1/4.

15. (a) The open circle in generation II has the genotype *–ee,* which indicates that each parent must carry an *e* allele. The brown pup in this generation is *bbE-,* indicating that each parent must also carry a *b* allele. Therefore, the genotypes are *BbEe* for both parents. (b) Based on similar logic as in part (a), the genotypes of II-1 and II-2 are *bbEe* and *Bbee* respectively. Therefore, for III-1, genotype *BbEe* has a probability of 2/3, and *BbEE* has a probability of 1/3. (c) One parent of III-7 is black and one is brown. Therefore, the *b* constitution of III-7 can be either *Bb* or *bb* each with the same probability. The gametes produced by III-7 are then *be* and *Be* in proportions 3/4 and 1/4, respectively. As in part (b) for III-1 we see that the genotype of III-4 is 2/3

BbEe and 1/3 *BbEE* The gametes produced, with the corresponding probabilities are 1/3 *BE*, 1/6 *Be*, 1/3 *bE*, and 1/6 *be*. A brown pup can have a genotype *bbEe* or *bbEE* but the latter is not possible here for only one parent carries *E* Therefore, the probability of a brown pup is the probability of gamete *be* (3/4) multiplied by that for *bE* (1/3), or 1/4 for black, 1/4 for splashed, and 1/2 for blue. The same answer is obtained by forming a Punnett square with the gametes and their appropriate probabilities and multiplying the values when forming zygotes.

16. The 1:2:1 ratio indicates that only one gene is involved; black and splashed are homozygotes and blue is the heterozygous phenotype. The probabilities of forming each of these phenotypes are 1/4, 1/2, and 1/4, respectively. There are six (3!) different orders in which exactly one of each type could be produced (for example, black, blue, splashed, or blue, black, splashed), so the probability of producing exactly one of each type is 6(1/4)(1/2)(1/4) = 3/16.

CHAPTER 2

Genes and Chromosomes

CHAPTER SUMMARY

A constant number of pairs of chromosomes is usually present in the somatic cell nuclei of higher plant or animal species. The two members of each pair are said to be homologous chromosomes, and each member of a pair is called a homologue. The homologous chromosomes of a particular pair have the identical appearance, but members of different pairs often have differences in size and structural detail that make them visibly distinct from each other. Cells with nuclei containing two sets of homologous chromosomes, one having come from the maternal parent and the other from the paternal parent of the organism, are said to be diploid. The gametes that unite at fertilization to produce the diploid somatic condition have nuclei with only one set of chromosomes, consisting of one member of each homologous pair; such cells are said to be haploid. Mitosis is the process of nuclear division that maintains the chromosome number when a somatic cell divides. Prior to mitosis, chromosomes replicate, forming a bipartite structure consisting of two chromatids joined at a region called the centromere. At the onset of mitosis chromosomes become visible and align along the midline of a structure called the spindle. For each chromosome the centromere divides and each half chromosome is presumably pulled by a spindle fiber to opposite poles of the cell. Later each set of chromosomes forms one of two daughter nuclei essentially identical to each other and to the initial nucleus. In the formation of germ cells the diploid number is reduced to the haploid number by meiosis. Replication of the genetic material occurs before the onset of meiosis, so every chromosome in each set consists of two chromatids. In the first meiotic division the sets separate and two cells result, each of which contains the replicated chromosomes, still consisting of two chromatids. In the second meiotic division, these chromatids separate. The two divisions of the nucleus and only one division of the chromosomes occurring in meiosis result in the separation, into different nuclei, of the tetrad of chromatids that make up a pair of homologous chromosomes, and provide a physical basis for the segregation and independent assortment of genes.

The chromosomes that control sex are commonly an exception to the rule that the chromosomes of a diploid organism consist of pairs of morphologically similar homologues. The sex chromosomes are called the X and Y in insects, mammals, and other organisms in which females are the homogametic (XX) and males the heterogametic (XY) sex. The Y chromosome carries either no genes or very few genes, depending on the organism. In mice and humans the few genes on the Y chromosome determine maleness. In *Drosophila* the ratio of the number of X chromosomes to the number of autosomes determines sex. The X chromosome carries many genes and is responsible for the association of certain phenotypes with sex; characteristics inherited in this way are said to be X-linked.

CHAPTER 2

Two methods that are particularly useful in the interpretation of genetic data are the binomial distribution and the chi-square method. The binomial distribution is useful in predicting the outcome of genetic events. The chi-square test is used to evaluate the fit between observed values and those predicted from a particular hypothesis.

BOLD TERMS

anaphase, anaphase I, anaphase II, autosomes, bivalent, cell cycle, centromere, chiasma, chiasmata, chi-square method, chromomeres, chromatids, chromosome, crossing over, degree of freedom, diakinesis, diploid, diplotene, G_1, G_2, haploid, hemizygous, hemophilia A, heterogametic, hermaphroditic, homogametic, homologous, interphase, intersexual, Klinefelter syndrome, leptotene, M, meiocytes, meiosis, metaphase, metaphase I, metaphase II, mitotic spindle, mitosis, nondisjunction, nucleoli, pachytene, Pascal's triangle, polar body, polyploidy, prophase , prophase I, prophase II, S period, second meiotic division, sex linked, sex chromosomes, spores, synapsis, synaptonemal complex, telophase, telophase I, telophase II, tetrad, Turner syndrome, wildtype, X chromosome, Y chromosome, zygotene.

ADDITIONAL INFORMATION

In mitosis, chromosomes divide once and the nucleus divides once, resulting in the formation of two cells identical to the parent cell and to each other, with respect to chromosome number. That is, if a diploid cell contains eight chromosome pairs (haploid number = 4), daughter cells produced by mitosis also contain eight chromosome pairs. Meiosis is a complex process, and it is frequently difficult to keep track of its many stages and to perceive its essential quality. Meiosis can be summarized as a process in which chromosomes divide once and the nucleus divides twice, resulting in the formation of four haploid nuclei from the original diploid parent cell. A distinctive feature of meiosis, and in fact the first feature that sets it apart from mitosis, is the side-by-side pairing (synapsis) of homologous chromosomes in the zygotene substage. During the diplotene substage the pairs become connected by chiasma and do not separate until anaphase I. This separation is called disjunction (unjoining) and failure to separate is called nondisjunction (no unjoining, an oft-confusing double negative). The distinction between mitosis and meiosis can best be seen by a side-by-side comparison, as shown in Figure 1. Note that (1) the chromosomes form different arrays in mitotic metaphase and meiotic metaphase I and (2) the second meiotic division is simply a mitotic sequence except that it is not preceded by chromosome replication, which occurs just before mitosis. The rare failures to separate homologues, that is, nondisjunction, occur during anaphase I. When this occurs, both homologues are present in anaphase II, so when the chromatids separate, each daughter cell has two copies of the chromosome that failed to separate. The probability of nondisjunction is so low that nondisjunction of two different

Figure 1 Comparison of meiosis and mitosis.

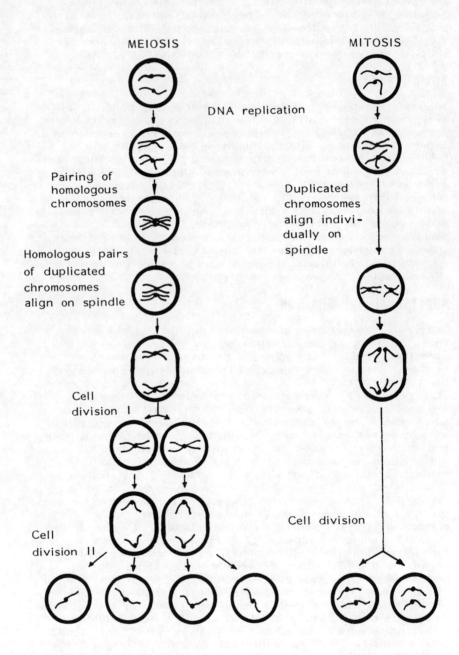

chromosomes is exceedingly rare indeed. When nondisjunction occurs with the X chromosome, a gamete forms that has a complete haploid set of autosomes but is diploid for the X.

Gamete production occurs by different mechanisms in animals and plants. In animals the meiotic products develop directly into gametes and the life cycle is entirely diploid except for the gametes. With few exceptions the sexes are separated into egg-producing females and sperm-producing males. The exceptions are the hermaphrodites (for example, certain worms), which produce both sperm and eggs. In plants the meiotic products are spores, which develop into a haploid gametophyte generation whose length varies widely from one species to the next. The gametophyte may exist as an independent small plant (as in the mosses) or may be contained in the mature plant (as in pollen and ovum production in higher plants). At any rate, the gametophyte produces the gametes. An outline of the life cycle of a plant is shown in Figure 2. In all higher plants the diploid generation is the major portion of the life cycle. In lower eukaryotes this is not the case. For example, in the yeast *Saccharomyces* both haploid and diploid cells can exist and multiply indefinitely. In contrast, in many algae the haploid stage predominates: the diploid zygote does not go through a mitotic stage but undergoes meiosis soon after fusion of male and female gametes.

Figure 2   Generalized life cycle of a higher plant

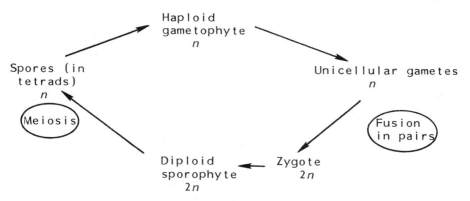

To understand the phenomenon of sex linkage, one must understand the concept of linkage and also know the sex-chromosome constitution of male and female organisms. The sex-chromosome makeup varies in different organisms, but in flies and humans a female is XX and a male is XY. If this had been understood in the past, fewer kings would have abandoned their queens, when the queen failed to produce a son. Since the queen could produce only X gametes and the king produced both X and Y, it is the king's Y-containing gamete that is needed, and the king would have to blame himself for failure to produce a son. (Probably, anyone telling a king that would have been beheaded!) It is important to realize that there is nothing special about sex linkage, for it is just another form of linkage. If two traits are determined by two genes carried on the same chromosome, these traits would not separate (segregate) during formation of gametes, as will be seen shortly. In the case of sex linkage one trait, sex, has two states (which is not different from most other traits) and femaleness is the phenotype of XX and maleness the phenotype of XY, in which the entire chromosome rather than a single gene is considered in evaluating the phenotype.

To clarify the effect of X linkage, we present the result of three crosses in which two genes are segregating; in one case, the genes are independent, in the second, they reside on the same chromosome, and in the third, one gene is carried on the X chromosome.

Cross I is between a red (Ww) and tall (Ss) doubly heterozygous female animal with a male having the identical genotype. The genes are on different chromosomes (they assort independently), neither of which are the X. Red (W) is dominant to white (w), and tall (S) is dominant to short (s).

Cross I $WwSs \times WwSs$

The result is the familiar 9:3:3:1 phenotypic ratio with equal numbers of males and females. That is,

9/32 red tall females
3/32 red short females
3/32 white tall females
1/32 white short females

9/32 red tall males
3/32 red short males
3/32 white tall males
1/32 white short males

In cross II the W and S genes are linked (this is designated by writing linked alleles on one side of a /); that is, they are on the same chromosome and do not assort independently. The parents are still heterozygous for each allele.

Cross II $WS/ws \times WS/ws$

In this case only two types of gametes are possible: WS and ws The genotypes are WS/WS, WS/ws, and ws/ws in a 1:2:1 ratio, which produce only two phenotypes in a 3:1 ratio, namely,

3/8 red tall females
1/8 white short females

3/8 red tall males
1/8 white short males

Note that linkage *decreases* the number of possible phenotypes.

In cross III, the gene for height is carried on the X.

Cross III $Ww\ X(S)X(s) \times Ww\ X(S)Y$.

As in cross I there are four gametes, $W\ X(S)$, $W\ X(s)$, $w\ X(S)$, and $w\ X(s)$ from the female. The male lacks the $w.X)s)$ gamete, having instead $w\ Y$. The Punnett square has the usual 16 entries (you may

CHAPTER 2   23

prepare it), but the number of phenotypes is less than that in cross I, because of the absence of certain phenotypes in one sex (you should check for yourself that the sex in which the particular phenotype is absent depends on whether the male lacks the dominant or recessive allele carried on the X). These phenotypes are

6/16 red tall females	3/16 red tall males
0 red short females	3/16 red short males
2/16 white tall females	1/16 white tall males
0 white short females	1/16 white short males

DRILL QUESTIONS

1. At what stage in the cell cycle does chromosome replication occur?

2. If a somatic cell contains 15 pairs of chromosomes just after completion of telophase, how many chromatids are present in early metaphase?

3. What structure in the chromosome is used to attach the chromosome to the spindle? What else is attached to this structure?

4. What is the name of the stage during which sister chromatids move to opposite poles of the spindle?

5. At what stage does chromosome replication occur in meiosis?

6. If a somatic cell contains 12 pairs of chromosomes, how many chromosomes are present per cell during telophase II? How many cells result from the complete series of meiotic steps from such a somatic cell?

7. If a somatic cell contains 12 pairs of chromosomes, how many visible chromosome units (bivalents) are present during pachytene?

8. At what stage do chiasmata become visible?

9. How many spindles are present in a tetrad in telophase II?

10. What is mean by a synaptonemal complex?

11. Are germ cells haploid or diploid?

12. What is meant by a sex-linked trait?

13. Distinguish the heterogametic sex from the homogametic sex.

14. How many different gametes, with respect to the sex chromosomes, can be produced by a human male?

15. What gametes can be produced by a human Aa male with respect to a gene not carried on the X?

16. What gametes can be produced by a human male with respect to a gene carried on the X? Consider the possibility of either *B* or *b* on the X.

17. If allele *A* is carried on the X and *B* is not, what gametes can be produced by a human female heterozygous for both loci?

18. A gene *D* is X-linked. Is it possible to perform a cross that might be written *Dd* x *DD*? Thnk about what is meant by this notation.

19. A gene *E* is X-linked. What is the genotype of all female progeny of a mating between a homozygous recessive female and a male with the dominant phenotype? Of all male progeny?

ANSWERS TO DRILL QUESTIONS

1. Interphase; that is, prior to prophase.

2. There are a total of 30 chromosomes in telophase. Since chromosome replication occurs in interphase, there are 60 chromatids present at metaphase.

3. The chromosomal component is the centromere; it is attached to a spindle fiber, which in turn is connected to a region near the poles of the spindle. Occasionally, the term kinetochore is used to mean the precise point of attachment in the centromere. In Chapter 5 you will learn about the structure of the centromere.

4. Anaphase.

5. As in mitosis, chromosome replication occurs during interphase, before meiosis begins.

6. Twelve chromosomes are present in each of four cells.

7. Because of the pairing of chromosomes in zygotene, the number is 12.

8. Diplotene.

9. Four.

10. This is a structure whose detailed features are not understood and that is present between chromosomes after synapsis. It is presumably responsible in some way for holding chromosomes together in register.

11. Haploid.

12. Any trait, one of whose determining genes is carried on the X chromosome.

13. The heterogametic sex possesses two different sex chromosomes. The homogametic sex possesses a pair of homologous sex-determining chromosomes. For example, in humans the male (XY) is heterogametic.

14. Two, one with an X and one with a Y. That is, there are X- and Y-containing sperm. Eggs are always X.

15. The segregation is no different from any other independently assorting genetic elements. Therefore, the gametes are XA, Xa, YA, and Ya. To indicate that the A gene is independent of the sex chromosome, one might write X/A, X/a, Y/A, and Y/a.

16. If B is on the X, the gametes are XB and Y-. If b is on the X, the gametes are Xb and Y-. To indicate that X and B are associated, one might write $X(B)$ and Y-, and $X(b)$ and Y-, for the two arrangements, respectively.

17. The genotype is $X(A)X(a)/Bb$. The four gametes are $X(A)B$, $X(a)B$, $X(A)b$, and $X(a)b$.

18. No, because it would imply a female mating with a female. One could however perform a cross of the type $DD \times D$-; DD would be the female and D- the male.

19. The female is ee and the male is E-. The progeny are Ee (dominant phenotype) females and e- (recessive phenotype) males.

ADDITIONAL PROBLEMS

1. A certain cell viewed at mitotic metaphase has two long chromosomes, three short chromosomes, and one medium-sized chromosome. Suggest an explanation for the chromosome sizes if the cell is diploid.

2. Meiosis can generate at least two different kinds of reorganization of genetic material. What are these two kinds?

3. How many chromatids and centromeres per cell will a diploid organisms with a haploid number of 26 have at the end of (a) metaphase I, (b) metaphase II, (c) telophase I, (d) telophase II, and (e) mitosis?

4. A normal woman whose father was a hemophiliac marries a normal man.
   (a) If this family has 4 children, 2 boys and 2 girls, what is the probability that (1) all 4 will be bleeders, (2) exactly three will be bleeders, (3) exactly 2 will be bleeders, (4) exactly 1 will be a bleeder, and (5) none will be bleeders?
   (b) If the couple had four children (sex unspecified), what is the probability that at least two will be bleeders?

5. In Martian goats long tails (T) are dominant to short tails (t). A farmer has noticed the two pedigrees shown in Figure 3.
   (a) What is the mode of inheritance of tail length?
   (b) Specify the distribution of individuals with long and short tails in the progeny of a mating between I-2 and II-6.

Figure 3

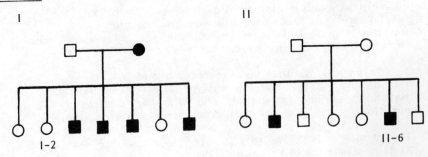

6. In *Drosophila,* brown eye color is determined by the recessive allele *bw* and miniature wing by the recessive allele *m*. A cross was made between a wildtype female and a male with brown eyes. Among the female progeny from the cross, 1/2 were wildtype and 1/2 were brown-eyed. The phenotypes of the male progeny were 1/4 wildtype, 1/4 brown-eyed, 1/4 miniature-winged, and 1/4 both brown-eyed and miniature-winged. Is either of the genes X linked? What are the genotypes of the parents?

7. In *Drosophila* two apparently wildtype flies produce the following offspring: 76 wildtype females, 26 ebony-bodied females, 38 wildtype males, 36 vermilion-eyed males, 12 ebony males, and 14 males with vermilion eyes and ebony body. What are the genotypes of the parent flies?

8. Assuming (1) that Turner females and Klinefelter males are equally viable, (2) that primary and secondary nondisjunctions are equally probable, and (3) that nondisjunction is equally probable in the male and the female (actually none of these statements are true), predict the relative frequencies of individuals with the Turner and Klinefelter syndromes in the population.

ANSWERS TO ADDITIONAL PROBLEMS

1. Note that the sizes do not all pair up. Clearly, the cell is from a male, and one short chromosome and the medium-sized chromosome are the sex chromosomes.

2. Meiosis can produce both intrachromosomal recombination by crossing over between homologous chromosomes and interchromosomal recombination by assortment of different chromosomes with respect to each other.

CHAPTER 2   27

3. (a) 104 chromatids, 52 centromeres. (b) 52 chromatids, 26 centromeres. (c) 52 chromatids, 26 centromeres. (d) 26 chromatids, 26 centromeres. (e) 52 chromatids, 52 centromeres.

4. (a) The woman's father was a hemophiliac; hence, she must have at least one h gene. It is very unlikely that she has more than one h gene since (i) the chance that her mother was a carrier is very slight and (ii) hh females tend to be nonviable in any event. Hence, the cross must have been $Hh \times HY$, producing progeny HH (normal female), Hh (carrier female), HY (normal male), and hY (hemophiliac male). Therefore, the following are the answers to the numerical problems: (1) All bleeders, 0. (2) Exactly 3 bleeders, 0. (3) The family has two boys. Since there is a 50 percent chance that any male offspring of that family will be a bleeder, the chance that both will be bleeders is $(1/2)(1/2) = 1/4$. (4) The boys could be born either in the sequence HY-hY or hY-HY. Since the probability of HY = that of hY = 1/2, the chance of exactly one bleeder is $(1/2)(1/2)2 = 1/2$. (5) The chance of no bleeders is 1 minus the probability of one or more bleeders, which has just been calculated; that is, $1 - (1/4 + 1/2) = 1/4$. (b) The probability that a child is a bleeder is equal to the chance that it is a boy multiplied by the probability that the boy has the h gene, or $(1/2)(1/2) = 1/4$. The chance that it is not a bleeder is $1 - 1/4 = 3/4$. The probability that at least two children will be bleeders is 1 minus the probability of no bleeders and one bleeder = $1 - (3/4)^4 + 4(1/4)(3/4)^3 = 67/256$.

5. (a) Since an affected mother transmits only to her sons (family I) and half of the sons of a phenotypically normal mother are affected, the trait is probably a sex-linked recessive. (b) Female I-2 has a long tail and is presumably Tt. Male II-6 has a short tail (t-). The mating will yield Tt (females), tt (females), T- (males), and t- (males), all in equal proportions. Thus, half of the progeny will have short tails.

6. The fact that both parents were wildtype suggests that both ebony (e) and vermilion (v) are recessive. Since ebony appears as the recessive homozygous class from a monohybrid class (a 3:1 ratio of wildtype to ebony in both males and females), both parents must be Ee. The sex differences in the distribution of eye color suggests X-linkage. Since vermilion appears in the male, the female must be Vv, but since it does not appear in the female, the male must not be heterozygous. Thus, the parental genotypes must be $VvEe$ (female) and EeV- (male).

7. One of the genes must be X-linked or the males and females would have the same phenotypic distribution. The female can be Bw Bw or Bw bw. If it were homozygous dominant, all offspring would have normal eyes. Thus, the female is heterozygous. For the same reason, the female must be

heterozygous for wing type and the male must be either heterozygous for wing shape or hemizygous (if it is X-linked). The male must carry the *bw* allele and can be either homozygous recessive or hemizygous. If it were hemizygous--that is, if the cross were *M m Bw bw* x *M m b-*, the male and female phenotypic distributions will be the same. Thus, the male cannot be *Mm b-*. Since one of the genes is X-linked, the *M-m* alleles are X-linked and the male must be *M-bw bw*. The Punnett square shows that this will yield the phenotypes given. Careful examination of the data will show that both genes could be X-linked

8. Turner females would be more frequent, since more nondisjunctions lead to XO individuals than XXY individuals. Figure 4 shows that nondisjunction in the female yields equal numbers of Klinefelter males and Turner females. On the other hand, Figure 5 shows that if all types of nondisjunction in the male are equally probable, Turner females and Klinefelter males would be produced in the ratio of 6:2. Thus, more Turner females are expected than Klinefelter males. (In fact, the opposite is true since more XO than XXY fetuses abort.)

Figure 4 Possible nondisjunctions in a female

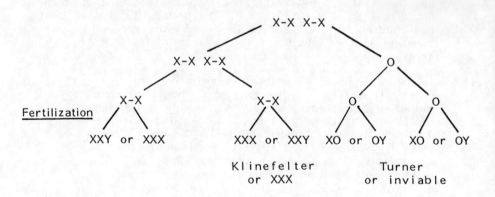

Figure 5 Possible nondisjunction in a male

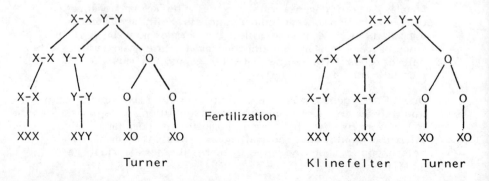

CHAPTER 2   29

SOLUTIONS TO PROBLEMS IN TEXT

1. (a) Deafness is confined to males, so it is probably
    sex-linked; cataract is found in both sexes, so it is probably
    autosomal. If deafness were dominant D, the mother in
    generation I would be dd and the father would be $d-$,
    and all males in generation II would be normal. Therefore,
    deafness is recessive d. The mother must be DD or
    Dd since deafness appears in generation II. If she were
    DD, no deafness would be present in generation II, which
    is not the case. Therefore, the mother is Dd, and the
    father is $D-$. Cataract cannot be recessive, because if so,
    the mating in generation II between two individuals with
    cataracts would yield only offspring with cataracts.
    Therefore, cataract is dominant (C). In that case, the
    original mother is cc and the father can be CC or
    Cc. The father cannot be CC, because all progeny in
    generation II would have cataracts, which is not the case.
    Therefore, the father in Cc. In summary, the mother was
    $Ddcc$ and the father was $D-Cc$.

2. The mother of the affected child must be heterozygous.
    Since the father must carry the dominant allele (since he is
    not affected), each daughter, for example, 1, could be either
    homozygous or heterozygous. Thus, denoting the recessive
    allele by a and the dominant by A, individual 1 has a
    probability of 1/2 of being Aa and 1/2 of being AA.
    Therefore, the probability of producing a gamete a is
    $(1/2)(1/2) = 1/4$. Individual 2 is $A-$ and produces half
    A gametes and half $-$ gametes. Thus, the probability of an
    affected offspring ($a-$) is $(1/4)(1/2) = 1/8$.

3. (a) Since the females, which have two X chromosomes, have
    three possible phenotypes, the genotypes and phenotypes are
    $X(c^b)X(c^b)$ (black), $X(c^b)X(c^y)$
    (calico), and $X(c^y)X(c^y)$ (yellow). Males can be
    either $X(c^b)Y$ (black) or $X(c^y)Y$ (yellow). In a
    cross between a black female (gamete $X(c^b)$ only) and a
    yellow male (gamete $X(c^y)$ or Y), the progeny will be
    have $X(c^b)X(c^y)$ (calico females) and half
    $X(c^b)Y$ (black males). (b) Since there are some yellow
    females ($X(c^y)X(c^y)$), both mother and father must
    carry the $X(c^y)$ allele, and the father must be
    $X(c^y)Y$ (yellow). Since there are black and calico
    progeny, the mother must also carry the $X(c^b)$ allele and
    hence be $X(c^b)X(c^y)$ (calico). (c) The male must
    have two X chromosomes to carry both alleles and still have a
    Y. The most likely genotype is $X(c^b)X(c^y)Y$, which
    is sterile.

4. (a) The probability of each child being a boy is 1/2. Thus,
    the probability of three boys of three children is
    $(1/2)(1/2)(1/2) = 1/8$. The probability of two families each
    with three children of identical sex is $(1/8)(1/8) = 1/64$. (b)
    There are two possible arrangements: A (all boys) and B (all
    girls) or A (all girls) and B (all boys); thus, $2(1/64) = 1/32$.

5. Note that the problem says nothing about the order of birth. Clearly, if the problem had stated that the first child was a girl, then the probability that the second is a girl will be 1/2. However, since two children have been born, the order of birth could have been BB, GG, GB, or BG. If one is a girl (that is, if the order were either GG, GB, or BG), then only one of the three possibilities just stated includes a girl as a second child. Therefore, the probability is 1/3.

6. The attached-X female produces the gametes $X(v)X(v)$ and Y, and the male produces $X(v^+)$ and Y gametes. Zygotes with an attached X and a normal X do not develop, so the only progeny are $X(v)X(v)Y$ (vermilion-eyed, attached-X females), and $X(v^+)Y$ (wildtype males).

7. The probability for each chromosome going to a particular pole is 1/2. There are 10 chromosomes, so the probabilities must be multiplied, and there are two possible pole. Thus, $(1/2)^{10} \times 2 = 1/512$.

8. (a) Let us call the recessive allele r. Since the son of II-1 is color blind, II-1 must be heterozygous (Rr). Since her husband is unaffected, he carries the dominant allele. Four types of progeny can result: RR females, Rr females, $R-$ males, and $r-$ males, or 3/4 unaffected. Thus, the probability of the next two children being normal is $(3/4)(3/4) = 9/16$. (b) Since there are color-blind individuals in generation II, the female in generation I must be Rr, and II-5 has a 50 percent probability of being Rr. Thus, her gametes and their probabilities are 3/4 R and 1/4 r. The male is affected ($r-$), so his gametes are 1/2 r and 1/2 Y-. The progeny genotypes and their probabilities are 3/8 Rr (normal), 3/8 $R-$ (normal), 1/8 rr (affected). and 1/8 $r-$ (affected). Therefore, the probability of the first child being color blind is $1/8 + 1/8 = 1/4$.

9. The fact that certain phenotypes are sex-associated indicates that the gene is probably carried on the X. Since there are three phenotypes, partial dominance is occurring and one can denote the possible genotypes of a female as NN, Nn, and nn, and those of a male by $N-$ or $n-$. Since all male progeny are normal, its mother must carry the normal allele N. Since progeny with intermediate eyes are all female, they must be Nn and the father must contribute the n allele. Since females with narrow eyes can be produced, both father and mother must carry the n allele. Therefore, the female is heterozygous (Nn), the male is hemizygous recessive ($n-$), and the gene for narrow eyes is recessive.

10. The female gametes are Bw and the male gametes are bW and $b-$. Therefore, the progeny are $WwBb$ (red-eyed, gray-bodied) females and $w-Bb$ (white-eyed, gray-bodied) males.

CHAPTER 2   31

11. The genotypes of the F_1 are *IiCcKk* (females) and *IiCcK-* (males). Work out a Punnett square remembering that only males and females can mate--that is, the cross is *IiCcKk* x *IiCcK-*. The female gametes (1/8 of each) are *ICK, ICk, iCK, iCk, IcK, Ick, icK,* and *ick;* the male gametes (1/8 of each) are *ICK, IC-, iCK, iC-, IcK, Ic-, icK,* and *ic-*. The result is 13/64 white fast males, 13/64 white slow males, 3/64 colored fast males, 3/64 colored slow males, and the same frequencies for the females.

12. (a) The probability of three boys and three girls is $(1/2)^6 = 1/64$, which must be multiplied by the number of ways three and three can be obtained from six children. This is obtained from Pascal's triangle, using row 6 and the coefficient for 3 and 3, namely, 20. Therefore, the probability is 20(1/64) = 20/64. (b) Here we have asked for a particular order, which can clearly be obtained in only one way. Therefore, 1/64(1) = 1/64. (c) Here we want the sum of the probabilities for 2 boys, 3 boys, ...6 boys, which can be obtained by multiplying 1/64 by the sum of the terms in row 6 of Pascal's triangle, starting with 2 boys and four girls--that is, (15 + 20 + 15 + 6 + 1)/64 = 57/64.

13. (a) Two heterozygous parents produce unaffected (U) and affected (A) children in the familiar 3:1 ratio. Therefore, the probability or four unaffected children is $(3/4)^4$ = 81/256. (b) The probabilities of affected and unaffected children are 1/4 and 3/4, respectively. Therefore, the probability of two affected children is the product (A x A x U x U) and the number of arrangements of the two among four children (6, obtained from Pascal's triangle), or (1/4)(3/4)(1/4)(3/4)6 = 27/128.

14. Sample 1, chi-squared = 6.66, df = 1, $P < 0.01$; sample 2, chi-squared = 1.32, df = 1, $0.3 < P < 0.5$. With larger sample size the same numerical deviation represents a smaller percentage error; larger samples provide a more critical test of a hypothesis, since proportionately larger deviations have a greater probability of occurring by chance in small samples.

15. Chi-square = 5.18, df = 3, $0.1 < P < 0.2$. Since deviations as great or greater would be expected to occur by chance alone in 10 to 20 percent of such samples, the data agree satisfactorily with a 9:3:3:1 ratio.

16. (a) $(8!/2!4!2!)(1/4)^2(1/2)^4(1/4)^2 = 105/1024$. (b) $(12!/6!2!3!1!)(9/16)^6(3/16)^2(3/16)^3(1/16)^1 = 0.0254$.

17. $7!/7^7 = 0.00612$.

CHAPTER 3

Gene Linkage and Chromosome Mapping

CHAPTER SUMMARY

Nonallelic genes located on the same chromosome tend to segregate together in meiosis rather than being independently assorted. This phenomenon is called linkage. The indication of linkage is deviation from the 1:1:1:1 ratio of phenotypes in the progeny of a cross of the form *AaBb* x *aabb* (a dihybrid testcross). Segregation of genes is not straightforward in a cross with two linked genes: more than 50 percent of the gametes produced by a dihybrid have parental combinations of the segregating alleles and fewer than 50 percent have nonparental (recombinant) combinations of the alleles. The recombination of linked genes occurs by crossing over, a process in which nonsister chromatids of the homologous chromosomes exchange corresponding segments during the first meiotic prophase.

The frequencies of crossing over between different gene loci can be used to determine the relative order and locations of the genes on chromosomes (genetic mapping). Distances between adjacent genes in such a map (a linkage map) is defined to be proportional to the frequency of recombination between them; the unit of map distance (the map unit) is defined as one percent recombination. One map unit corresponds to a physical length of the chromosome in which a crossover event will occur, on the average, in one of every 50 meioses. The recombination frequency underestimates actual genetic distance if the interval between the genes being considered is long. This discrepancy results from the occurrence of multiple crossover events, which yield either no recombinants or the same number produced by a single event. For example, two crossovers in the interval between two genes will not be evident by the production of recombinants, and three crossover events will yield recombinants of the same type as that produced by a single crossover.

When many genes are mapped in a particular species, they are found to occur in linkage groups equal in number to the haploid chromosome number of the species. The maximum frequency of recombination that can occur between any two genes in a cross is 50 percent; this frequency arises when the genes are on nonhomologous chromosomes and assort independently, or when the genes are sufficiently far apart on the same chromosome that at least one crossover events occurs between them in every meiosis.

The analysis of linkage and recombination using the four haploid products of individual meiotic divisions is possible in some species of fungi and unicellular algae. The method is called tetrad analysis. In *Neurospora* and related fungi the meiotic tetrads are contained in a tubular sac or ascus in a linear order, making it possible to determine whether the segregation of a pair of alleles occurred in the first or the second meiotic division. With such asci it is possible to use the centromere as a genetic marker, and in fact the centromere serves as a reference point to which all genes in the same chromosome can be mapped.

CHAPTER 3   35

BOLD TERMS

asci, ascospore, ascus, attached-X chromosome, chromosome map, *cis* configuration, *cis*-heterozygote, *cis-trans* test, coefficient of coincidence, complementation, complementation groups, compound-X chromosome, coupling, crossing over, first-division segregation, genetic map, heterokaryon, interference, linkage, linkage group, linkage map, locus, map unit, mapping function, nonparental ditype, parasexual cycle, parental ditype, recombinant, recombination, repulsion, second-division segregation, spores, tetrad, tetratype, three-point crosses, *trans* configuration, *trans*-heterozygote.

ADDITIONAL INFORMATION

The general topic of linkage and recombination is strictly based on logic and probabilities, but nonetheless it is often difficult for the beginner. In this section some of the features that frequently cause trouble are reviewed.
    The principle of genetic mapping is straightforward and based on the relation between map distance and the probability of exchange. For fairly short distances mapping is done quite directly. For example, the following genetic distances lead to a simple map: *a-b* 1, *b-c* 2, *a-c* 3. The map is *a*-1-*b*-2-*c*, and there is no ambiguity because 1 + 2 = 3. In actuality, the sum is rarely so perfect in a particular experiment because of the statistical nature of genetic exchange. For example, in five different experiments, the observed values of *a-b* might be 1.2, 0.85, 1.16, 0.93, and 0.86, which would average to 1.0. If map distances are larger, they are usually not strictly additive. For example, the segment of the map *d*-12-*e*-14-*f* might yield a value of *d-f* of 21 rather than 12 + 14 = 26. The most usual reason for such nonadditivity is the increasing probability of double crossovers with increasing distance. That is, the frequency of recombinants is not directly proportional to the frequency of crossing over, because between two markers double crossovers, which do not lead to recombination, go undetected. This problem is discussed in the text; for the purpose of this review, the message is simply that a value for the map distance between *d* and *f* is obtained more reliably by adding the distances *d-e* and *e-f;* this method is optimal because direct determination of *d-f* will surely underestimate the real value. In fact, values such as 12 and 14 map units may already be sufficiently large that they may too be underestimates of the distances *d-e* and *e-f*. At any rate, adding *d-e* and *e-f* will give a value more nearly correct than the direct determination of *e-f;* furthermore, if map distances for even smaller intervals were available, it would be best to sum these instead. In general, when constructing maps, the shortest intervals should be considered as the most reliable and should be used first. Problem 6 in the Additional Problems section provides an example of this reasoning. The student

should be warned though that in systems in which very small distances can be measured, for example, bacteriophages, in which a recombination frequency of 0.0001 can be determined, distances are rarely additive for these very short distances. The reason is that one crossover may enhance or inhibit the occurrence of a second crossover in the immediate vicinity; this topic, which is called high negative interference, will not be discussed.

It should be noted that a genetic map may be more than 50 units long even though the maximum recombination frequency never exceeds 50 percent. This is because distances much less than the length of a single chromosome may be large enough that the probability of crossing over is nearly 1. The following map is an example of these phenomena (Figure 1) Note the various regions in which distances are not additive. In particular, note that the interval $a-p$ has a recombination frequency of 43 percent, yet by adding the individual intervals, the map length is 96 units long. It should be realized that as more mapping data are obtained, the entire length of a map can grow. For example, additional markers between h and j might change this region of the map to $h-6-s-9-q-6-r-9-t-5-u-3-j$, which would mean that $h-j$ is 38 rather than the 32 shown on the larger map, increasing the total length of the map to 102. The map length of bacteriophage T4 grew over a period of a few years from 300 units to more than 1500 map units as additional markers were mapped.

Figure 1 A hypothetical genetic map.

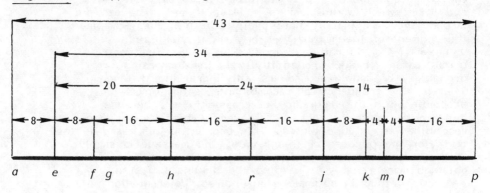

Two precautions must be taken in calculating recombination frequencies from raw data. One is simply the necessity of counting all recombinations whether they are single crossovers or part of a double-crossover event. Consider a cross $AbD \times aBd$ (gene order $A.B.D$) yielding the results: 729 AbD and 739 aBd (parental combinations), 160 Abd and 148 aBD (single crossover in the $B-D$ region), 110 ABd and 98 abD (single crossover in the $A-B$ region), and 9 ABD and 7 abd (double crossover). The total number of progeny is 2000. In calculating the map distance between A and B the single exchanges in A-B are not sufficient because each double crossover represents an additional exchange in that region. Therefore, the recombination frequency is (110 + 98 + 9 + 7)/2000 = 11.2 percent, rather than 10.4 percent, the value that would result if the double crossovers were neglected. Similarly, the recombination frequency for the region B-D is (160 + 148 + 9 + 7)/2000 = 16.2 percent. Note that the double-crossover class is counted twice, once for each interval.

The second precaution is necessary when studying recombination in viral systems. When performing a cross such as $Ab \times aB$ with viruses, it is not unusual that the parental types and the AB recombinant can be detected easily, but that the double mutant ab is either not detectable easily or not at all. The problem becomes quite severe with multiply mutant viruses. It cannot be overemphasized that measuring recombination frequency requires that both AB and ab recombinants be counted. However, because recombination is almost always reciprocal, the number of ab doubly mutant recombinants can be taken to be equal to the number of AB wildtype recombinants. Thus, recombination frequency can be estimated as 2(number of AB recombinants)/total. For example, in a phage cross one might obtain 962 Ab, 914 aB, and 62 AB, for a total number of detected phage of 1938. To calculate the recombination frequency, we add 62 ab (that is, a number equal to the number of AB) both to the number of recombinants and the total detected phage, so the recombination frequency is 2(62)/(1938 + 62) = 124/2000 = 6.2 percent.

Genes can clearly be ordered by pairwise mapping and combining map distance, as just described. However, as we have pointed out in the text, a three-factor cross enables three genes to be mapped in a single cross. Although this is adequately described in the text, it is worthwhile to remind the student of the essential features of this procedure. Two types of information are obtained by analysis of the data from three-factor crosses--the gene order and the distances--and these types are obtained from different components of the data. The gene order is obtained simply from identifying the two parental combinations of genes (the largest numbers) and the double-crossover classes (the smallest numbers). In the double-crossover recombinants the central markers are exchanged with respect to the parental combinations, so gene order can be determined immediately. Recombination values for the single-crossover classes are not needed for ordering the genes. However, these values provide the recombination frequencies and hence the map distances for the intervals in which the crossing over has occurred.

The analysis of tetrads often causes confusion for beginners. In principle, it is straightforward, but the analysis appears complex primarily because of the use of terms describing segregation patterns to refer to events that occurred during meiosis. The high points will be reviewed briefly. The logic of mapping from order-tetrad data is the following. First, one uses the value of the number of asci in which second-division segregation has occurred for a particular gene to determine the distance between that gene and the centromere. Each gene is examined separately. One then looks at the relative frequencies of parental-ditype and nonparental-ditype tetrads. If they are equal, the genes are the known to be unlinked, and no further mapping is required: that is, the distance from each gene to a centromere is known, but one does not know whether the genes are on different chromosomes or on the same chromosome and so far apart that they are unlinked. If the number of nonparental-ditype asci is very small compared to the parental-ditype asci, the genes are the known to be linked and hence on the same chromosome. In that case, there is reason to continue the mapping process. As described in the text, it is possible to determine the relative order of linked genes with respect to the centromere without examining map distances. One simply looks at whether first- and second-division segregation of two genes is present in a single type of ascus. For example, consider the order centromere-A-B. With that order, all asci in which second-division segregation occurs for A, must also exhibit second-division segregation for B, because a crossover between the centromere and A is also between the centromere and B. In addition, there must be asci showing second-division segregation for B and not for A, because a crossover can occur between A and B. If the genes are on opposite sides of the centromere, that is, if the order is A-centromere-B, most asci will show second-division segregation for either A or B, but very rarely for both, which would require two crossovers. To determine distance between a gene and the centromere, from ordered-tetrad data, requires a measure of the recombination frequency, and this can be obtained by counting the asci in various classes. However, remember that the definition of recombination frequency is the fraction of all progeny that are recombinant and note that if a crossover occurs, a tetrad will consist of two recombinant chromosomes and two nonrecombinant chromosomes. Therefore, if asci with recombinants (second-division segregations) are counted and compared to the number of all asci, one must divide by 2, because there are eight times as many progeny as total asci but only four times as many recombinant progeny as asci with second-division segregations. The recombination frequency for each marker will yield the distance between each gene and its centromere and thus the distance between the genes is simply the difference between the distances. If the distances centromere-A and centromere-B are large, the difference A-B is obtained by subtracting two large numbers, which is statistically inaccurate. In that case, recombination frequencies can be obtained directly by examining tetratype asci.

In unordered asci, it is not possible to identify first-division and second-division segregation by examination of the spores in the ascus, since these patterns differ only in their orders. Thus, one cannot map markers directly with respect to the centromere, and the relative positions of markers is obtained from examination of the tetratype asci. There is a cumbersome method for mapping a gene with respect to a centromere from unordered-tetrad data, but it will not be described here. Usually, if a gene is sought that is very closely linked to the centromere, and once found it can be used as a reference marker for the mapping of other genes with respect to the centromere. This technique is commonly used in the genetics of yeast, which produces four-spore asci.

Our final review concerns complementation analysis. The principle reason for carrying out a complementation test is to determine whether two mutations are allelic, that is, if they are in the same gene. This information is often used to make a count of the minimum number of genes responsible for a particular phenotype. For example, if a collection of mutations that determine a particular phenotype fall into six complementation groups, one can usually conclude reliably that at least six genes contribute to the phenotype. The number is always a minimum value, because mutations may not yet have been isolated in other genes. In rare cases, a gene might also be missed if mutations in that gene are lethal. Classifying mutations in different complementation groups is straightforward. One combines the mutations in a single organism, but on different chromosomes and examines the phenotype. The rule is simple: if the mutations are in different genes, a diploid individual carrying both mutations will contain one good copy and one bad copy of each gene product and the two good copies will usually give the individual the wildtype phenotype. An example with three genes, A, B, and C, should illustrate how to count genes. Assume that each gene is represented by a single mutant allele a, b, and c, that capital letters are dominant, and that the wildtype phenotype requires a genotype with at least one copy of each capital letter. We start by crossing a mutant $AAbbCC$ with another mutant $aaBBCC$. The F_1 must be $AaBbCC$, and it will have the wildtype phenotype. One says that a and b complement one another. The same result will be obtained by crossing $AABBcc$ and $AAbbCC$ (b and c complement), and $aaBBCC$ and $AABBcc$ (a and c complement). The crosses would be summarized in Figure 2.

Figure 2

3 genes	a	b	c
a	−	+	+
b	+	−	+
c	+	+	−

Suppose a and b were mutations in the same gene. Then, the first cross would have been $aaCC \times bbCC$ and the F_1 would be $abCC$ (a and b do not complement), which would have a mutant phenotype. The other crosses would have been $bbCC \times AAcc$, yielding the wildtype $AbCc$ (b and c complement), and $aaCC \times AAcc$, yielding the wildtype $AaCc$ (a and c complement). This would be summarized in the table in Figure 3.

Figure 3

2 genes	a	b	c
a	−	−	+
b	−	−	+
c	+	+	−

This example is an unfair one since we knew the answer at the outset. Suppose we had the data and had to deduce the answer. In the three-gene case, it would have been obvious that there were three genes, because no row would have a − entry other than in a self cross; that is, every mutation would complement every other mutation. In the two-gene case, one observes that row a and row b each have two − entries, indicating that a and b do not complement. Thus, a and b are allelic. Mutation c complements both a and b and is clearly in a separate gene.

DRILL QUESTIONS

1. What gametes are produced by an individual whose genotype is $AaBb$ if the genes are (a) on different chromosomes or (b) on the same chromosomes, assuming no crossing over?

2. What gametes are produced by an individual whose genotype is AB/ab, if crossing over can occur between the genes?

3. What are the genotypes and phenotypes of the progeny of a mating $AB/ab \times Ab/aB$? Assume no crossing over.

4. Which of the following arrays is a *cis* configuration: ab/AB or aB/Ab?

5. What gametes are produced by male and female *Drosophila* having the genotypes AB/ab, assuming that crossing over occurs?

6. If all the chromosomes in an organism were mapped, how many linkage groups would be found in (1) a haploid organism with

17 chromosomes per somatic cell, (2) a virus with only one chromosome (which is typical for viruses), and (3) a rat, with 42 chromosomes per somatic cell?

7. For closely linked genes the frequency of recombinant progeny is half the frequency with which chiasmata occur between their loci. Why half?

8. Does the statement in Problem 7 hold for all genes?

9. A cross is carried out between AB/aB and ab/ab. The following progeny and the numbers of each produced: 215 $AaBb$, 209 $aabb$, 36 $Aabb$, and 39 $aaBb$. Which progeny are recombinants and what is the recombination frequency between A and B?

10. The recombination frequency between genes A and B is 6.2 percent. By how many map units are the genes separated in the linkage map?

11. A cross $ABD/abd \times abd/abd$ is performed. If the map order is alphabetical, which of the following progeny is a nonrecombinant, and which is the result of a single or a double crossover: ABD/abd, abd/abd, ABd/abd, Abd/abd, abD/abd, AbD/abd, aBd/abd?

12. The recombination frequency between Q and R is 3.7 percent. Does it matter whether the map is represented $Q-3.7-R$ or $R-3.7-Q$?

13. The recombination frequencies between a, b, and c are: ab, 1 percent; ac, 2 percent; and bc, 3 percent. What is the genetic map of these genes?

14. Given the recombination frequencies between genes a and b, a and c, and c and d, how many more values are needed to construct a map?

15. A three-factor cross $ABD/abd \times abd/abd$ is carried out which yields the fact that the recombinant gamete AbD is formed at a much higher frequency than Abd. What is the gene order? Assume equal spacing of the genes.

16. A cross $A \times a$ occurs in *Neurospora* in which an exchange occurs between the gene and the centromere. What is the order of the genotypes of the spores in the resulting ascus?

17. A cross $Abd \times aBD$ is performed in *Neurospora*. If exchange occurs between genes B and D, what is the order of the genotypes of the spores in the ordered ascus?

18. There are eight spores in a *Neurospora crassa* ascus. Since there are only four meiotic products, either the first or last division in the ascus must be mitotic. What is the evidence that it is the last?

19. In a cross *ABD/abd* x *abd/abd*, what are the genotypes of the double-crossover classes, if the map order is *A.D.B*?

20. In a cross *ABD/abd* x *abd/abd* the most common progeny are *ABD/abd* and *abd/abd*, and the least frequent are *aBD/abd* and *Abd/abd*. What is the gene order?

21. In a cross *ABD/abd* x *abd/abd* in which the gene order is *B A D*, which single crossover classes will be represented by nearly the same number of progeny?

22. The genetic map for three genes is *A*-8-*B*-12-*D*. In a cross *ABD/abd* x *abd/abd* the frequency of *aBd* is 0.0072. What is the coefficient of coincidence?

23. An *A/a* heterokaryon is formed in *Neurospora* and the resulting asci are studied. If no crossing over occurs between the centromere and the gene, what is the arrangement of the spores in the ascus and is this a first- or second-division segregation pattern?

24. If a single crossover occurs between a gene and a centromere in *Neurospora*, why is the spore pattern called a second-division segregation pattern? That is, why does segregation not occur in the first division?

25. An *AB/ab* heterokaryon is formed in *Neurospora*. If no crossing over occurs, a parental ditype ascus forms. What is the gene order in such an ascus?

26. If *AB/ab* heterokaryons are formed in *Neurospora*, why are nonparental ditype asci so infrequent compared to parental ditypes? How is the frequency of tetratype asci compared to that of nonparental ditype asci?

27. Two mutations, *a* and *b*, are known to produce a white coat in homozygous individuals (*aa* or *bb*) of the normally black *Paracelsus hellenica*. The mutations are linked. The genotype *a/b* has the wildtype phenotype. Are the mutations in the same gene?

28. Three mutations *1*, *2*, and *3* produce the wildtype phenotype in the *trans* configuration in the combinations *1-2* and *1-3*, but not in the combination *2-3*. What are the complementation groups?

ANSWERS TO DRILL QUESTIONS

1. (a) *AB, Ab, aB, ab*. (b) *AB* and *ab*, or *Ab* and *aB*, depending on which alleles are linked.

2. The noncrossover gametes are *AB* and *ab*; the crossover gametes are *Ab* and *aB*.

CHAPTER 3   43

3. The gametes of one parent are *AB* and *ab;* those of
    the other parent are *Ab* and a*B*. The genotypes are
    AB/Ab, AB/aB, ab/Ab, and *ab/aB,* in equal numbers.
    The phenotypes are *AB, Ab,* and *aB,* in the ratio 2:1:1.

4. The *cis* configuration is *ab/AB.*

5. The female will produce *AB, ab, Ab,* and *aB.* The
    male will produce only *AB* and *ab,* because crossing
    over does not occur in *Drosophila* males.

6. (1) 17; (2) 1; (3) there is one linkage group per
    chromosome type and rats are diploid (have two copies of each
    chromosomes), so the number of linkage groups is (1/2)42 = 21.

7. Each recombination event is preceded by a chiasma. Since
    only two of the four chromatids present in a tetrad
    participate in forming a particular chiasma, each chiasma
    results in only two of the four chromatids being recombinant.
    Hence, only half of the four meiotic products would be
    crossover types and the frequency of recombinant offspring
    would be half the frequency of chiasmata.

8. No. When genes are closely linked, it is rare for more than
    one chiasma to occur between them, and the statement is true.
    However, when the genes are far apart, multiple crossover
    events occur, and the probability that two crossovers may
    cancel each other affects the validity of the statement.

9. The nonrecombinant gametes are *AB* and *ab* from one
    parent and *ab* from the other parent, yielding *AaBb* and
    aabb progeny. Recombinant gametes are *Ab* and *aB*
    yielding the progeny *Aabb* and *aaBb.* The recombination
    frequency is the sum of the number of recombinants (36 + 39 =
    75) divided by the total number of progeny (499), or 15
    percent.

10. 6.2 map units.

11. The gametes from one parent are *ABD* and *abd;* the
    other parent produces only the gamete *abd.* Therefore, the
    parental zygotes are only *ABD/abd* and *abd/abd,*
    and *AaBbDd* is the only nonrecombinant type. The singly
    recombinant gametes are *ABd* and *abD* (arising from a
    single crossover betwen *B* and *D*) and *Abd* and
    aBD (arising from a single crossover *A* and *B*).
    These four gametes yield the recombinant types *AaBbdd,
    aabbDd, Aabbdd,* and *aaBbDd,* respectively. The genotypes
    aaBbdd and *AabbDd* are derived from the doubly
    recombinant gametes *aBd* and *AbD,* respectively.

12. No, unless a third gene has been given as a reference.

13. *b*-1-*a*-2-*c*.

14. Two more: for example, one to locate *c* to the right or left of *a* and another to locate *d* to the right or left of *c*.

15. This can be worked out most easily by writing down the cross with the three possible gene orders: (I) *A B D*, (II) *A D B*, and (III) *B A D*. Recalling that single crossovers are much more probable than double crossovers, we can predict that for I *bAD* occurs much less frequently than *Abd*, for II *AbD* occurs with comparable frequency with *Abd*, and that for III *AbD* occurs much more frequently than *Abd*. Hence only order III agrees with the data.

16. This is second-division segregation. Four orders are possible: *AAaaAAaa*, *aaAAaaAA*, *AAaaaaAA*, and *aaAAAAaa*.

17. | *Abd* | *Abd* | *AbD* | *AbD* | *aBd* | *aBd* | *aBD* | *aBD*; |
    | *AbD* | *AbD* | *Abd* | *Abd* | *aBd* | *aBd* | *aBD* | *aBD*; |
    | *Abd* | *Abd* | *AbD* | *AbD* | *aBD* | *aBD* | *aBd* | *aBd*; or |
    | *AbD* | *AbD* | *Abd* | *Abd* | *aBD* | *aBD* | *aBd* | *aBd*. |

18. There are eight spores in the sac, but only four products of meiosis. Therefore, either the zygote or the spores must undergo mitosis. However, almost invariably the eight spores are found as four pairs of identical spores (exceptions to this rule will be discussed in Chapter 8), which could occur only if the last division were mitotic.

19. Remember that the central marker will be exchanged. Thus, the genotypes are *ABd/abd* and *abD/abd*.

20. The most frequent classes indicate the noncrossover classes. The least frequent are the double crossovers, in which the central markers are exchanged. Thus, the central markers are *A* and *a*, so the gene order is *B A D*. Note that this is the same order as *D A B*; left and right are completely arbitrary in a genetic map.

21. Crossing over between *B* and *A* yields the progeny *Bad* and *bAD* in roughly equal numbers; crossing over between *A* and *D* yields equal numbers of *BAd* and *baD*.

22. The genotype shown results from a double crossover. The expected frequency is $(0.08)(0.12) = 0.0096$. The coefficient of coincidence is $0.0072/0.0096 = 0.75$.

23. The spore arrangement is *AAAAaaaa*; it is first-division segregation, for *A* and *a* separate in the first meiotic division.

24. If there is no exchange, *A* and *a* separate from one another when the *A/A/a/a* tetrad divides in meiosis I. When an exchange occurs, meiosis I yields two chromosomes with the composition *A/a*; in other words, *A* and *a* have not yet separated from one another. In the second meiotic division each *A/a* unit separates yielding one *A* spore and one *a* spore; thus, segregation occurs in the second division.

25. *AB,AB,AB,AB,ab,ab,ab,ab*.

26. Nonparental ditype asci are rare because they result from a four-strand double crossover, whereas parental ditype asci form when no crossovers occur. Tetratype asci occur more frequently than nonparental ditypes because tetratypes are formed by a single crossover, or a 3-strand crossover.

27. No. If they were, diploid cells would have two defective genes that determine the same trait.

28. Since *1* and *2* complement, they are in different groups. Similarly, *1* and *3* are in different groups. However, *2* and *3* fail to complement, so they are in the same group. Thus, there are two groups: one contains *1* and the other contains *2* and *3*.

ADDITIONAL PROBLEMS

1. Compare the phenotypes produced in a cross between *Ab/aB Drosophila* males and *ab/ab* females versus *Ab/aB* females and *ab/ab* males. Assume that the genes are on an autosome.

2. In *Potabila aqua* a dark tail (*D*) is dominant to a clear tail (*d*) and a short tail (*S*) is dominant to long tail (*s*). The genes are linked. In a cross between one homozygous recessive parent and another whose genotype is *DS/ds*, the following progeny result: 2198 dark short, 2219 clear long, 68 dark long, and 75 clear short. What is the recombination frequency between the tail-length and tail-color genes?

3. A cross *gB/Gb* x *gb/gb* is carried out with the following results: 351 parental types and 645 recombinant types. These results are reported in a scientific paper submitted to a reputable journal, but the paper is immediately rejected with the comment that "something is clearly wrong with the data." Why should this simple experiment have elicited such a reaction?

4. Consider Problem 3 again. Suppose you repeat the experiment several times and always obtain the same results. Can you suggest any explanation for the data? (Think about the difference between production of a gamete and production of an organism.)

5. A cross *AaBbDd* x *aabbdd* is carried out with the following results: *AaBbDd* 47; *AaBbDd* 207; *aaBbDd* 43; *aabbDd* 205; *AaBbdd* 201; *Aabbdd* 43; *aaBbdd* 207; *aabbdd* 47. Which genes are linked, and what is the distance between the linked genes?

6. Construct a map of a chromosome from the following map distances between individual pairs of genes: *r-c* 10,

c-p 12; p-r 3; s-c 16; s-r 8. You will
discover that the distances are not strictly additive. Think
of some way to deal with the small discrepancies. It is not
necessary to calculate the frequencies of double crossovers.

7. In *Drosophila* the cross a +/+ b x a +/+ b
always leads to a 2:1:1:0 ratio. This fact is frequently used
by geneticists when studying a newly isolated mutant. How do
you think they use it?

8. In *Neurospora* 10 percent of the tetrads show
second-division segregation for *arg,* and 0 percent for
pab. If *arg* and *pab* assort independently, what is
the expected ratio of PD, NPD, and TT tetrads in a cross
arg pab x + +?

9. Three mutations in *Neurospora*--*a, b,* and *c,* are on
the same side of the centromere and are 5, 11, and 18 map
units from it, respectively. A cross abc x +++ is
performed.
(a) What fraction of the asci will show second-division
segregation for *b*?
(b) What fraction of the spores will show a recombination
event between *a* and *c*?
(c) What fraction of the recombinant asci (between *a*
and *c*) will show second-division segregation of *c*?
(d) What fraction of the recombinant asci will show
second-division segregation for *c* but not for *b*?
(e) What fraction of the total asci will show
second-division segregation for *b* but not for *a* or *c*?

ANSWERS TO ADDITIONAL PROBLEMS

1. The problem is concerned with the curious phenomenon
that crossing over does not occur in *Drosophila*
males. Therefore, when the male is *Ab/aB,* his
gametes are only *Ab* and *aB;* thus, the progeny are
Aabb and *aaBb* (two genotypes, two phenotypes).
In contrast, when the female is *Ab/aB,* her gametes
are both the parental types, *Ab* and *aB* and the
recombinant types *AB* and *ab.* Therefore, the progeny
are *Aabb, aaBb, AaBb,* and *aabb* (four genotypes,
four phenotypes.

2. Dark long and clear short are recombinants. Therefore, the
recombination frequency is 68 + 75 = 143 divided by the total
number of progeny (4560), or 3.1 percent.

3. Note that the recombination frequency is greater than 50
percent, which is certainly grounds for rejection unless the
data are somehow justified.

4. One factor that must be considered in any cross is whether

the probabilities of survival of all possible genotypes are the same. For example, a wildtype organism often has a greater survival than one with many (say, 20) mutations. In fact, examples have already been given of lethal combinations of alleles. If recombinant genotypes have a higher probability of survival than parental genotypes, a recombination frequency greater than 50 percent may be observed. This effect may be occurring in this cross, in which one of the recombinants is wildtype. Examination of the relative numbers of the two recombinant types would be informative. For example, if wildtype had greater viability than the double mutant, there might have been 510 GB/gb and 135 gb/gb. The inequality in the number of the two recombinant types would be suspicious and make one suspect that the wildtype recombinant not only has greater viability than the homozygous recessive recombinant, but also greater than the parental types. Another possibility, which is rare but quite interesting, is that four-strand double-crossovers occur very frequently. These crossovers yield four recombinant products (you should check this) and occur more frequently than two-strand double crossovers, which yield no recombinant products. This is one situation in which the observed recombination frequency is greater than 50 percent.

5. It is necessary to examine the genes in pairs to answer this question. The rule is simple: unlinked pairs will be present in equal numbers in all combination, whereas linked pairs will not. For example, a cross $F/f;G/g$ x $f/f;g/g$, in which a / separates homologues and a ; separates unlinked chromosomes, will yield equal numbers of FG, Fg, fG, and fG progeny. In contrast, the cross FG/fg x fg/fg will yield more FG and fg (parental) offspring than Fg and fG (recombinant) offspring. The data are rearranged by counting the number of types of each pairs of alleles, as shown below:

AB	248	AD	254	BD	90
Ab	250	Ad	244	Bd	408
aB	250	aD	248	bD	412
ab	252	ad	254	bd	90
	1000		1000		1000

For the $A-B$ pairs and for the $A-D$ pairs, all four combinations occur with equal frequency, indicating that A and B are unlinked, and that A and D are unlinked. The combinations of $B-D$ pairs occur with unequal frequencies, so B and D must be linked. The genotypes of the $B-D$ pairs, indicate that the genotype of the heterozygous parent is $A/a;Bd/bD$. The recombination frequency for the $B-D$ pair is $(90 + 90)/1000 = 0.18$, so the B and D genes are about 18 map units apart.

6. The smaller numbers are likely to be the more accurate

because of the relatively small number of double exchanges. Thus, assume that the smaller values (3, 5, 8) are the more accurate, arrange them first, and then place the other markers in a way that minimizes discrepancies. The map is
s-5-p-3-r-10-c.

7. Since it is only true if a and b are linked, it is a straightforward criterion for linkage between a known and an unknown marker.

8. Since *arg* and *pab* assort independently, they are not linked, and are either on separate chromosomes or more than 50 map units from one another on the same chromosome. Since *pab* shows no second-division segregation, it is located right next to its centromere. Since *arg* shows 10 percent second-division segregation, it must be 5 map units from a second centromere (a different centromere since the genes are not linked). Thus, the fraction of tetrads that are tetratypes is determined by the number of recombination events between each marker and its centromere, which is 10 percent in this case. Since there is no linkage between the genes, the remaining 90 percent of the tetrads will be divided equally between PD's and NPD's. Hence, the expected ratio is 9 PD : 9 NPD : 2 TT.

9. (a) Twice the recombination frequency or $(2)(0.11) = 0.22$. (b) 0.18 will show a recombination event between c and the centromere, and 0.05 will show recombination between a and the centromere. Thus, $0.18 - 0.05 = 0.13$ will show a recombination event between a and c. (c) All except those with double crossovers (between a and the centromere and a and c). Thus, $1.00 - (0.05)(0.13) = 0.994$. (d) All recombinants except those with double crossovers (between a and the centromere and b and c). That is, $1.00 - (0.05)(0.07) = 0.997$. (e) These will be the asci in which a double crossover has occurred, one between a and b, and another between b and c, or $(2)(0.06)(0.07) = 0.0084$.

SOLUTIONS TO PROBLEMS IN TEXT

1. The genes are X-linked, so the father contributes no genes to male offspring. The normal son can only be derived from a ++ gamete produced by the mother, and such a gamete requires recombination between the *gd* and *hemA* genes. The recombination frequency between these genes is 16 percent. This value is for production of both recombinant progeny, namely, ++ and *gd hemA*. Therefore, the ++ gamete will arise at half that value, or 0.08.

2. One parent is *Gl ra/gl Ra*, and the other is *gl ra/gl ra*. The number of recombinant types is $6 + 3 = 9$. The total number of progeny is $88 + 6 + 103 + 3 = 200$. Therefore, the recombination frequency is $9/200 = 4.5$ percent.

CHAPTER 3   49

3. There is no recombination in the male, so the male produces gametes ++ and $b\ cn$, each with a probability of 0.5. The recombination frequency is 0.09, which means that 0.09 of the gametes are recombinant and 0.91 are parental; thus, the probabilities of the different gametes are 0.455 ++, 0.455 $b\ cn$, 0.045 + cn, and 0.045 b +. The genotypes of each progeny type is obtained from a Punnett square, and the frequencies are obtained by multiplying the probabilities. The results are 0.2275 $b\ b\ cn\ cn$, 0.455 $b + cn +$, 0.2275 + + + +, 0.0225 $b\ b\ cn$ +, 0.0225 $b + + +$, 0.0225 $b + cn\ cn$, and 0.0225 + + cn +.

4. First, the genes are not assorting independently, so they must be linked. The first cross is $RRPP \times rrpp$. The genotype of the F_1 is $RrPp$. The nonrecombinant gametes are RP and rp; the recombinant gametes are Rp and rP. The gametes of the homozygous recessive of the testcross are all rp. Thus, the progeny of the testcross are $RrPp$ (dark-eyed), $rrpp$ (light-eyed), $Rrpp$ (light-eyed), and $rrPp$ (light-eyed). Since there were 628 dark-eyed rats, there must have been about 628 light-eyed nonrecombinant $rrpp$ individuals. Therefore, light-eyed products of recombinant gametes must number 889 − 628 = 261. Thus, the recombination frequency is 261/(628 + 889) = 0.172. In the second cross the F_1 are all Rp/rP. The nonrecombinant gametes are Rp and rP; recombinant gametes are rp and RP. The testcross yields $Rrpp$ (light-eyed) and $rrPp$ (light-eyed) nonrecombinant types and $rrpp$ (light-eyed) and $RrPp$ (light-eyed) recombinant types. Since 86 are dark-eyed, there must be 2 × 86 = 172 recombinant types. Therefore, the recombinant frequency is 172/(86 + 771) = 0.201. It is not at all uncommon for repeated experiments or crosses carried out in different ways to yield slightly different values for the recombinant frequency--for example, 0.17 and 0.20. A more reliable value is obtained by averaging the experimental values, that is, $(1/2)(0.172 + 0.201) = 0.186$.

5. The major classes, which in a testcross must represent the allelic constitution of the chromosomes of the heterozygous parent, are colored nonwaxy shrunken (C c Wx wx sh sh) and colorless waxy plump (c c wx wx Sh sh). Therefore, the genotype of the heterozygous parent is C Wx sh/c wx Sh. Now consider the three possible map orders with either Wx, Sh, or C as the central gene, and note which genotypes would result from two crossovers. The correct order must yield colorless waxy shrunken (2 progeny) and colored nonwaxy plump (4 progeny) as the minority classes. Only the order with Sh in the middle agrees with this observation. Thus, the map is C Sh Wx. Map distances can now be determined from the number of recombinants resulting from a single crossover. A crossover between C and Sh yields colored plump waxy (113 progeny) and colorless shrunken nonwaxy (116 progeny). Thus, the recombination frequency is (113 + 116 + the double crossovers)/6708 = 0.035. A crossover

between Sh and Wx yields colored shrunken waxy (601) and colorless plump nonwaxy (626), yielding a recombination frequency of (601 + 626 + the double crossovers)/6708 = 0.184. Therefore, the genetic map is c-3.5-sh-18.4-wx.

6. Homozygous recessive individuals arise when a double crossover occurs that produces a triply recessive gamete. The map distances between Sh and Wx are 59 - 29 = 30 map units and between Wx and Gl 10 map units. Therefore, the expected frequency of double crossovers is (0.3)(0.1) = 0.03. Since both triply recessive and triply dominant gametes are produced together at this frequency, the probability of the triple recessive is half that value or 0.015. The expected number of homozygous recessive individuals among 2000 progeny is (0.015)2000 = 30. The coefficient of coincidence is the ratio of the observed number to the expected number; since this is 0.5, the observed number is (0.5)30 = 15.

7. (a) When one parent is homozygous recessive, the largest classes indicate the genotypes of the chromosomes of the other parent. Thus, the parent is $K\ Cd.\ e/k\ cd\ E$. There are three possible gene orders, with either $Cd, K,$ or e in the middle. The minority classes indicate the gametes formed by a double crossover. Only the gene order with E in the middle is consistent with the data. The recombination frequencies are then determined from the single-crossover classes. Exchange between K and E yield + + cd (58) and $k\ e\ Cd$ (67) and a recombination frequency of (58 + 67 + the double-crossover classes)/2000 = 0.066. For exchanges between E and Cd, the frequency is (49 + 44 + 3 + 4)/2000 = 0.05. Thus, the map is k-6.6-e-5.0-cd. (b) If crossing over occurred independently, the frequency would be (0.066)(0.005) = 0.0033, or (0.0033)2000 = 6.6 double-crossover progeny. The observed number is 7.

8. (a) Recombination between S and ar is 0.57 percent; recombination between S and dp = 10.9 percent; recombination between ar and dp = 10.9 percent. The possible orders are $S\ ar\ dp$, $S\ dp\ ar$, and $ar\ S\ dp$. With either order having S at the end, crossing over between it and the nearest of the other markers would yield some wildtype and some triple mutant progeny. Therefore, the correct order is the third one. The genetic map is ar-0.57-s-11.0-dp. (b) The very short interval between ar and S results in a frequency of double crossovers so low that double-crossover progeny will usually not be found in a sample of 2118 progeny. Wildtype and Star aristaless dumpy, representing the double-crossover classes, are missing.

9. Garnet and vermilion are linked, 11.2 map units apart. Orange-eyed males are double-mutant recombinants. The two phenotypic classes in the F_1 indicate that the genes are X-linked; that is, the parents were vG/vG females and

Vg/Y males. The F_1 wildtype females were vG/Vg and vermilion males were vG/Y. The recombination frequency is 11.2 percent, and the recombinant chromosomes yield wildtype males and some of the wildtype females, plus the orange-eyed males in which the chromosome constitution is vg/Y.

10. The 1766 asci represent those in which crossing over has not occurred either between *a* and *b* or between the centromere and either of the genes. The 220 asci are those in which a crossover has occurred between the centromere and the nearest of the two genes. If a crossover occurred between *a* and *b*, the order of the spore pairs would be a^+b^+ a^+b ab^+ ab (for the order centromere *a* *b*) or a^+b^+ ab^+ a^+b ab (for the order centromere *b* *a*. Thus, the latter is the correct order. The recombination frequency between the centromere and *b* is $(1/2)220/(1766 + 220 + 14) = 0.055$; and between *b* and *a* is $(1/2)14/2000 = 0.035$. Therefore, the map is centromere-5.5-*b*-0.35-*a*.

11. The equivalence of parental-ditype and nonparental-ditype asci means that the loci are segregating independently: they are either on nonhomologous chromosomes or more than 50 map units apart on the same chromosome. In other words, the genes are unlinked.

12. (a) Distance from the centromere to a gene is calculated from the number of second-division segregations (s) by the equation

$$\text{Map distance} = \left(\frac{1}{2}\right)\frac{100s}{\text{Total tetrads}}$$

Page 89 of the text lists the four marker arrangement for second-division segregation. Thus, the distance from *m* to the centromere is $(1/2)(76 + 4 + 5)100/400 = 10.6$ map units, and the distance from *t* to the centromere is $(1/2)(54 + 4 + 5)100/400 = 7.9$ map units. Therefore, the arrangement of the genes is *m*-10.6-centromere-7.9-*t*. (b) $(1/2)(76 + 4 + 54 + 1 + 5)/400 = 0.17$. (c) An entire ascus is scored as showing a second-division segregation, but only half of the spores contain a chromosome with a crossover in the region of interest.

13. The parents are *a*++ and +*bc*. If all were linked, most tetrads would consist of 2 *a*++ and 2 +*bc*, which is not the case. If none were linked, all combinations would be equally frequent, which is also not the case. Note in rows 1 and 2 that there are equal numbers of *b* and +, assorting at random with the *a*+ and +*c* combination; similarly, comparison of rows 3 and 4, shows that *b* assorts independently of the *a* and *c* genes. Therefore, *a* and *c* are linked, and *b* is unlinked to *a* or *c*. Rows 3 and 4 have the recombinant arrays *ac* and ++. The recombination frequency is $(1/2)(62 + 74 + 1 + 1)/410 = 17.3$ percent.

14. (a) Segregation can occur in either the first or second division. Let us use p as the frequency of a second-division segregation of one gene and $1 - p$ as the frequency of first-division segregation. Similarly, we use q and $1 - q$ for the frequencies of second- and first-division segregation for the other gene. Tetratypes can arise with segregation at the first or the second division, with a frequency $q(1 - p) + p(1 - q)$. They can also occur if there is both first- and second-division segregations, but in only half (1/4 lead to PD and 1/4 to NPD--work this out yourself), that is, at a frequency of $pq/2$. Therefore, the total frequency of TT is $q(1 - p) + p(1 - q) + pq/2 = p + q - 3pq/2$.
In this example, the recombination frequencies are 0.05 and 0.10, so $p = 0.10$ and $q = 0.20$. Therefore, TT $= 0.10 + 0.20 - 3(0.10)(0.20)/2 = 0.27$. NPD = PD and NPD + PD + TT = 1, so NPD = PD = 0.365. (b) NPD + TT/2 = 50 percent.

15. One must first determine the gametes and the frequency of each. It is best to begin with the double crossovers, so that the single crossover values can be corrected from the number of map units. Double-crossover classes would occur at a frequency of $(1/2)(0.05)(0.10) = 0.0025$ each. However, the coefficient of coincidence is 0.6, so the observed values will be 0.0015 each. Crossovers between Pg and Gl will occur at a frequency of 0.05 - the double-crossover classes, or 0.0485 each. In the region between Gl and Bk, the observed frequencies will be 0.25 - 0.0015 = 0.0235. The parental classes will be 1 - (the sum of all recombinant classes), or 0.4265 each. Multiplying these frequencies by 2000 yields the number of individuals of each type. The parental classes are 853 dark green, glossy, nonbrittle and 853 pale green, nonglossy, brittle; crossovers between Pg and Gl yield 47 dark green, nonglossy, brittle and 47 pale green, glossy, nonbrittle; crossovers between Gl and Bk yield 97 dark green, glossy, brittle and 97 pale green, nonglossy, nonbrittle; the double crossover classes are 3 dark green, nonglossy, nonbrittle (wildtype) and 3 pale green, glossy, brittle.

16. In calculating a map distance single crossovers must be corrected for double crossovers. That is, the total distance must equal the sum of the individual frequencies minus the double-crossover frequency (x), or $(0.15 - x) + (0.20 - x) = 0.34$. Solving for x yields a frequency of double crossing over of 0.005.

17. The recombinant gametes and the frequency of each produced by each parent are $(1/2)P$ dpy unc and $(1/2)P$ ++. Dumpy uncoordinated progeny result only by fertilization of a dpy unc egg with a dpy unc sperm, which would occur at a frequency of $(1/2)P(1/2)P = (1/4)P^2$.

18. (a) Since fpa, leu, and $ribo$ are associated in both classes, it is likely that they are on the same chromosome;

similarly, *w* and *ad* are associated and *ar* probably
on a second chromosome. (b) Remember that the selective marker
(*fpa*) is always selected to be near the tip of the
chromosome on the same side of the centromere as the genes to
be mapped. Thus, the possible gene orders are
centromere-*leu-ribo-fpa* or centromere-*ribo-leu-fpa*.
With the first order, the *fpa* single crossovers will yield
leu$^+$ribo and *leu$^+$ribo$^+$*, and rare
ribo$^+$leu by double crossing over. The second order
will yield *ribo$^+$leu* and *ribo$^+$leu$^+$* by
single crossovers, and rare *leu$^+$ribo* by double
crossing over. The observed collection includes
leu$^+$ribo, which are not rare, and
ribo$^+$leu, expected to be rare, are absent.
Therefore, the order is centromere-*leu-ribo*.

19. Remember how an allele is defined: as an alternate form of a gene. If two mutations are noncomplementing, then they are judged to be mutations in the same gene and thus are alleles; if they complement, then they are in different genes and are not alleles.

20. To distinguish complementation and noncomplementation by examination of phenotype, the heterozygous genotype of each gene must have the wildtype phenotype, when mutations are in different genes. For example, if dominant mutants were used, the phenotype of a heterozygote would depend only on the presence of the particular mutation.

21. Consider row 1: there are - entries with 3 and 7, so 1, 3, and 7 form one group. Row 2: all +, so 2 is the only member of a group. Row 3: mutation already classified. Row 4: there is a - for 6 only and no other - in column 4, so 4 and 6 form a group. Row 5: - only with 9 and no other - in column 5, so 5 and 9 form a group. Rows 6 and 7: mutations already classified. Row 8: all +, so 8 is the only member of a group. In summary, there are five complementation groups: (1) 1, 3, 7; (2) 2; (3) 4, 6; (4) 5, 9; (5) 8.

22. Consider the first cross. We arbitrarily set *sn* and *v* to the left and right, respectively, of *lz^{46}*. Then *lz^k* can be either to the left or right of *lz^{46}*. If it were to the left of *lz^{46}*, the v-extreme lz phenotype would arise by a single crossover. A triple crossover would be necessary if *lz^k* were to the right of *lz^{46}*. This is consistent with forming the sn-normal phenotype by a single crossover. Now consider the second cross in which we determine whether *lz^k* is the left or right of *lz^{BS}*. If *lz^k* were to the right of *lz^{BS}*, a single crossover would give extreme lz, as observed. With *lz^k* to the right of *lz^{BS}* the sn-extreme lz is produced by a double crossover in which the relatively large space between *sn* and *lz^{BS}* is available for the exchange. Combining the results of the two crosses yields *sn lz^{BS} lz^k lz^{46} v*.

23. Mutation 9 is in complementation group B but the mutant gene product inhibits the activity of the gene product of a wildtype copy of gene B. In each cell in which B should be fully active, the + response is weak, because the total activity of the inhibited B is not adequate for a normal + response. Mutation 9 is said to be anticomplementing.

CHAPTER 4

The Chemical Nature and Replication of the Genetic Material

CHAPTER SUMMARY

Except for certain viruses, the genetic material of all organisms is DNA. In prokaryotes the DNA is dispersed throughout the cell. In eukaryotes the major fraction is enclosed in the nuclear membrane; some DNA is also found in cytoplasmic organelles, such as mitochondria and chloroplasts. DNA was shown to be the genetic material, in contrast with proteins, through (1) the phenotypic transformation of bacteria by isolated DNA and (2) the injection of phage DNA, but not phage protein, into bacteria that subsequently produce progeny phage. DNA is a double-stranded polymer consisting of repeating nucleotides. A nucleotide has three components--a base, a sugar (which is deoxyribose in DNA), and a phosphate. Sugars and phosphates alternate in forming a single polynucleotide chain that has distinct termini, a 3'-hydroxyl group and a 5'-phosphate group. In double-stranded DNA the two strands are antiparallel: each end of the double helix carries one 3'-OH group and one 5'-P group. Four bases are found in DNA--adenine and guanine (purines) and cytosine and thymine (pyrimidines). Equal numbers of purines and pyrimidines are found in double-stranded DNA. The bases are paired; only pairs between adenine and thymine (an AT pair) and guanine and cytosine (a GC pair) are present. This pairing is responsible for holding the two polynucleotide strands together in a double helix. The base composition of DNA varies from one organism to the next. The information content of a DNA molecule lay in the sequence of bases along the chain, and each gene consists of a unique sequence.

The double helix replicates using enzymes called DNA polymerases, but other proteins are also needed. Replication is semiconservative in that each parental single strand, called a template strand, is found in one of the double-stranded progeny molecules. Replication proceeds by a DNA polymerase (1) bringing in a nucleoside triphosphate capable of hydrogen-bonding with a base in a template strand, (2) removing the terminal diphosphate from the nucleoside triphosphate, and (3) joining the 5'-P group of the nucleoside monophosphate to a free 3'-OH group of the growing strand. The polymerases can only join a 3'-OH group and a 5'-P group. Since double-stranded DNA is antiparallel, only one strand--the leading strand--grows in the direction of movement of the replication fork. The other strand, the lagging strand, is synthesized in the opposite direction as short fragments that are subsequently joined together; the overall direction of joining of the fragments is in the direction of movement of the replication fork. DNA polymerases cannot initiate synthesis, so a primer is always needed. The primer is an RNA fragment made by an RNA-polymerizing enzyme; the RNA primer is removed at later stages of replication. The leading strand and each fragment of the lagging strand are initiated by RNA primers. DNA molecules can be both linear and circular. In both cases, replication is usually bidirectional. A DNA molecule of a prokaryote usually has a single replication loop; in contrast, eukaryotic DNA molecules usually have many loops. Circular molecules replicate in one of two ways: in the Θ form or by rolling circle replication. The latter generates a linear branch, which may be either single- or double-stranded. In the Θ mode DNA gyrases are needed to prevent tangling of the unreplicated portion. The base sequence of a DNA molecule can be determined by one of several methods. With each method, the DNA is broken into discrete fragments containing 100-200 nucleotide pairs. Complementary strands of each fragment are sequenced. The sets of overlapping fragments have been combined to determine the complete sequence of several phage chromosomes.

BOLD TERMS

active site, adenine (A), agarose, antiparallel, autoradiography, bacteriophages, base composition, base stacking, clone, cohesive ends, covalently closed circles, complementary, cytosine, DNA gyrase, DNA ligase, DNA polymerase, DNA polymerase I, DNA polymerase III, DNA-unwinding proteins, editing function, equilibrium centrifugation in a density gradient, endonuclease, eukaryote, exonuclease, gel electrophoresis, genetic code, guanine (G), gyrase, helicase, hydrophobic interaction, *in vitro* experiment, initiation, lagging strand, leading strand, Maxam-Gilbert sequencing method, nucleic acid, nucleoid, nucleoside, nucleotide, 3'-OH group, Okazaki fragment, 5'-P group, partial denaturation mapping, percentage G+C, phages, phosphodiester bond, polynucleotide chain, precursor fragment, primase, primer, prokaryote, proofreading, purine, pyrimidine, renaturation, replication fork, replicative form, replication origin, RF, ribose, RNA polymerase, rolling circle replication, semiconservative replication, single-stranded DNA-binding protein, substrate, tautomeric shift, template, template chain, thymine, topoisomerase, transformed, uracil, virus.

ADDITIONAL INFORMATION

The material in this chapter is sufficiently straightforward that there seems to be no reason for going over it again. It almost seems that if you remember that A = T and G = C, you have the whole story. Instead, a few bits of information will be provided that enlarge slightly on some of the material.
    Several arguments were given for the advantage of a double-stranded structure and these not been reviewed. Chemically, the structure also provides protection for the base sequence. The DNA bases are hydrophobic and hence form a stacked array within the center of the double helix. In fact, the bases are so near one another that water is excluded. Since most chemical reactions within cells take place in aqueous solution, the net effect of the water exclusion is that the bases are protected from being attacked by many reactive chemicals. Interestingly, many partially hydrophobic reactive molecules are mutagenic and carcinogenic; the probable explanation is that their slightly polar components

enable them to dissolve in water but their large hydrophobic (and, incidently, reactive) component enables them to "go into solution" in the hydrophobic core of DNA.

In our discussion of DNA synthesis several accessory proteins were mentioned--helicases, single-stranded DNA-binding (ssb) proteins, and gyrases--but their functions were not described. DNA polymerase I and DNA polymerase III differ significantly in their ability to polymerize different types of molecules. For example, *in vitro* a DNA molecule with one single-strand break with a 3'-OH group serves as a site of addition of nucleotide addition by polymerase I but not by polymerase III. When polymerase I works at such a site, it either removes the deoxynucleotides in its path by means of its 5'-3' exonuclease activity (the one needed to remove RNA from precursor fragments prior to joining) or displaces the DNA ahead of it, producing a single-stranded "tail." This reaction is called strand displacement and is indicative of the ability of polymerase I to unwind the helix. Polymerase III fails to carry out such a reaction, because it is unable to unwind DNA. *In vivo,* DNA polymerase III always follows behind a helicase, an enzyme whose function is to unwind double-stranded DNA. Because of the precursor-fragment mechanism, synthesis of the lagging strand is always delayed with respect to the leading strand, and hence single-stranded DNA is always present in the replication fork. The single-stranded segment is the unreplicated portion of the template for the lagging strand. Intramolecular hydrogen bonds can form between bases within a single strand, so this single-stranded DNA can collapse on itself as a tangle of partially double-stranded segments. Such chaos is avoided by the ssb protein, a protein that binds to the single-stranded DNA. In fact, as the helicase unwinds DNA ahead of the replication fork, ssb protein immediately binds to the DNA to prevent formation of double-stranded DNA. Polymerase III is able to displace the ssb protein, and replication proceeds.

DRILL QUESTIONS

1. Name the bases in DNA and in RNA.

2. Which sugars are in DNA and in RNA, and how do they differ?

3. Which bases pair in DNA and form intrastrand pairs in RNA?

4. What is a nucleoside, and how do a nucleoside and a nucleotide differ?

5. How many phosphate groups are there per base in DNA and in RNA, and how many phosphates are there in each precursor for DNA and RNA synthesis?

6. What chemical groups are at the ends of a single polynucleotide strand?

7. What is the relation between the amount of DNA in a somatic cell and in a gamete?

CHAPTER 4   59

8. Name four requirements for initiation of DNA synthesis.

9. What is mean by base stacking?

10. To what chemical group in a DNA chain is an incoming nucleotide added and what group in the nucleotide reacts with the DNA terminus?

11. By what radioisotopes can DNA and protein be distinguished?

12. In what sense are the two strands of DNA antiparallel?

13. What is the base sequence of a DNA strand that is complementary to the hexanucleotide 3'-A-G-G-C-T-C-5'? Label the termini.

14. What is meant by a nuclease, and how do endonucleases and exonucleases differ?

15. In what direction does a DNA polymerase move along a template strand?

16. How do organisms solve the problem that all DNA polymerases move in the same direction along a template strand, yet double-stranded DNA is antiparallel?

17. What is the chemical difference between the groups joined by a DNA polymerase and by DNA ligase?

18. Name three enzymatic activities of DNA polymerase I.

19. Why can RNA polymerases and primases initiate DNA replication whereas DNA polymerases cannot?

20. What must be done to two precursor fragments before they can be joined together?

21. How does rolling circle replication differ from theta replication with respect to the templates that are used?

22. In gel electrophoresis do smaller double-stranded molecules move more slowly or more rapidly than larger molecules?

ANSWERS TO DRILL QUESTIONS

1. DNA: adenine, guanine, cytosine, thymine. RNA: adenine, guanine, cytosine, uracil.

2. DNA: deoxyribose. RNA: ribose. Ribose has an OH group on the 2' carbon and deoxyribose has an H.

3. DNA: A and T, G and C. RNA: A and U, G and C.

4. A nucleoside is a base attached to a sugar. A nucleotide is

a nucleoside phosphate.

5. Both DNA and RNA: one P per base. Both DNA and RNA: three phosphates in each precursor.

6. A 3'-OH group and a 5'-P group.

7. A somatic cell, which is diploid, has twice as much DNA as a gamete, which is haploid, except for the small difference due to the different sizes of the sex chromosomes.

8. A template, a primer, a DNA polymerase, and four nucleoside triphosphates.

9. Back stacking is the tendency for bases in DNA or RNA to be oriented such that their planes are parallel and so that one base is directly over or nearly over the other in the chain.

10. The 5'-P of the incoming nucleotide reacts with the 3'-OH of the primer.

11. Phosphorus is found in DNA but not proteins, and sulfur is present in proteins, but not nucleic acids.

12. At one end of double-stranded DNA one strand terminates with a 3'-OH group and the other strand with a 5'-P group.

13. 3'-G-A-G-C-C-A-5'.

14. From the 3' end to the 5' end.

15. A nuclease is an enzyme capable of breaking a phosphodiester bond. An endonuclease can break any phosphodiester bond, whereas an exonuclease can only remove a terminal nucleotide.

16. One strand is copied from the 3' end to the 5' end, and the other strand is copied in the direction opposite to that of the movement of the replication fork by synthesis in short pieces.

17. DNA polymerases join a 5'-triphosphate to a 3'-OH group and in so doing remove two phosphates so that the phosphodiester bond contains one phosphate; a ligase joins a single 5'-P to a 3'-OH group.

18. Polymerizing activity, 5'-3' exonuclease, 3'-5' exonuclease.

19. Neither RNA polymerases nor primases require a primer.

20. The RNA primer must be removed from the precursor that was made first, because DNA ligase cannot join DNA to RNA.

21. In rolling circle replication only one strand of the parental circular molecules serves as a template; the second

strand of the branch is copied from a progeny strand. In theta replication both parental strands are templates.

22. Smaller molecules move faster, because they can penetrate the pores of the gel more easily.

ADDITIONAL PROBLEMS

1. A culture of bacteria grown for a very long time in ^{14}N-containing medium is transferred to ^{15}N-containing medium for one-half generation. What is the density distribution if the DNA is isolated either in the standard way in the form of fragments having a molecular weight of 5-10 million, or as an intact molecule?

2. Consider a hypothetical phage whose DNA replicates exclusively by rolling circle replication. A phage whose DNA is radioactive in both strands infects a bacterium and is allowed to replicate in nonradioactive medium. Assuming that only daughter DNA from the branch ever gets packaged into progeny phage particles,
(a) What fraction of the parental radioactivity will appear in progeny phage?
(b) How many progeny phage will contain radioactive DNA? Will the occurrence of crossing over affect this value?

3. How long will it take to replicate a medium-sized phage DNA molecule whose molecular weight is 25 million (a typical size) and which replicates bidirectionally?

4. In the interest of economy, cells do not usually make more of any enzyme than is needed. What do you think would be the relation between the number of DNA polymerase III molecules in E. coli and the corresponding polymerase in animal cells?

5. Compare the rates of migration of a circular DNA molecule and its linearized form (perhaps broken by a nuclease) in gel electrophoresis.

6. A technique is used for determining base composition. Rather than giving mole fractions for individual bases, it yields the value of [A]/[C]. If this value is 1/3, what fraction of the bases are A?

7. E. coli DNA has a molecular weight of about 2.7×10^9. How long is this molecule and what does this tell you about the state of the intracellular DNA? (The length of a typical rodlike bacterium is 10^{-4} cm.)

8. Bases other than the standard four bases are found in cells. Several of these, for example, hypoxanthine, are able to base-pair with one of the four standard bases. However, these base are never incorporated into DNA. Suggest a few possible explanations.

9. If DNA polymerase I is supplied with dATP and dTTP, but no primer and no template, DNA containing dA and dT will actually form. Interestingly, the base sequence is that of strict alternation of A and T, that is, ATATATATAT... The reaction is exceedingly slow, taking many hours before DNA appears. However, once DNA is detected, the amount then begins to increase very rapidly. Explain this rapid increase.

10. What is the fundamental difference between the initiation of theta replication and of rolling circle replication?

ANSWERS TO ADDITIONAL PROBLEMS

1. After one-half generation half of the DNA is unreplicated and will still have the original density. The other half has replicated, yielding hybrid DNA, but the mass of the replicated material is twice that of the unreplicated region. Therefore, if fragmented, 2/3 of the DNA will be hybrid and 1/3 will have the parental density. If unfragmented, there will be a single band whose density is the average of 2/3 hybrid and 1/3 parental.

2. (a) In rolling circle replication one parental strand remains in the circular portion and the other is at the terminus of the branch. Therefore, only half of the parental radioactivity can appear in progeny.
(b) If there is no crossing over between progeny DNA molecules, one progeny phage will be radioactive. If crossing over occurs once, two particles will be radioactive. If crossing over is frequent and occurs at random, the radioactivity will be distributed among the progeny.

3. The molecular weight of a nucleotide pair is about 660, so the DNA molecule contains 38,000 nucleotide pairs. The replication rate of *E. coli* DNA is 50,000 nucleotide pairs per minute per replication fork. Thus, the time to replicate the phage DNA is $(1/2)(38,000/50,000) = 0.38$ min, or 22 sec.

4. Since *E. coli* has two active replication forks and eukaryotic cells have thousands, one would expect animal cells to have several thousand times as many DNA polymerase molecules as a bacterial cell.

5. The circle is more compact and therefore will move more rapidly.

6. If $[A]/[C] = 1/3$, then $([A] + [T])/[G] + [C]) = 1/3$, and the DNA is 25 percent A+T, or 12.5 percent A.

7. From the molecular weight of a nucleotide pairs (660) the number of pairs is about 4×10^6. The spacing between nucleotide pairs is 3.4 A = 3.4×10^{-8} cm, so the length is $(4 \times 10^6)(3.4 \times 10^{-8}) = 0.136$ cm.
The DNA is circular so if stretched out, it would be $(1/2)1360 = 680$ times the length of the bacterium. Thus, the DNA must be highly folded in the cell.

CHAPTER 4   63

8. Remember that enzymes are highly specific. The enzymes that form nucleosides and nucleotides may not recognize these bases. If nucleotides can form, the DNA polymerases may not recognize them as DNA precursors.

9. Obviously the enzyme has great difficulty forming any DNA at all, so a long time elapses before any DNA forms. However, once formed, it can fold back on itself and form a double-stranded hairpin, because of the alternating ATAT... sequence. This double-stranded DNA provides a template. Some of the short dinucleotides that form in the early stages of template-free synthesis can serve as a primer. Thus, as soon as some DNA is made, the rate of synthesis begins to approach that for normal template-mediated replication.

10. Rolling circle replication must be initiated by a single-strand break. Theta replication does not need such a break.

SOLUTIONS TO PROBLEMS IN TEXT

1. DNA: thymine; RNA: uracil. Actually, thymine is found in some tRNA molecules, though it is formed from uracil after the tRNA is synthesized. Also, there are a few known phages that have uracil in their DNA instead of thymine; modified forms of uracil, for example, hydroxyuracil, have also been observed.

2. A nucleotide contains a phosphate group. Remember, a nucleotide is a nucleoside phosphate.

3. (a) If 18 percent is adenine, 18 percent is also thymine, for a total of 36 percent for A + T. Therefore, G + C will be 64 percent, and C will be half that, or 32 percent. (b) ([G] + [C])/[all bases]) = 0.64 AT.

4. (a) The distance between nucleotide pairs is 3.4 angstrom units or 0.34 nm. Thus, the total number of nucleotide pairs is $(34 \times 10^{-6})/(3.4 \times 10^{-10}) = 10^5$. (b) There are ten nucleotide pairs per turn of the helix, so $10^5/10 = 10^4$.

5. (a) The base composition is not meaningful in terms of the standard definition because of the lack of complementarity. Therefore, the individual percentages would normally be stated. (b) Clearly, [A] is not equal to [T], nor is [G] equal to [C]. Thus, the DNA cannot be double-stranded and is probably single-stranded.

6. Because of the strict alternation of A and T, any region of the molecule can base-pair with any other. Thus, the molecule will fold back onto itself and form a hairpinlike structure. Of course, many different hairpinlike structures are possible, depending on the particular bases that pair. Interestingly, each structure fluctuates because of breathing, with base pairs breaking and re-forming, but not necessarily with pairing between the bases that separate. Because long tracts

of AT pairs break easily at room temperature, short hairpin molecules exhibit the phenomenon of slippage--that is, the relative lengths of the two sides of the hairpin continually change as if the two parts were sliding back and forth along one another.

7. (a) The pairing is much like the random assortment of gametes in a mating between two heterozygotes, that is a 1:2:1 ratio. Thus, there are three bands, in the ratios 1 $^{15}N^{15}N$: 2 $^{14}N^{15}N$: 1 $^{14}N^{14}N$.
(b) T4 and P2 DNA have no parts of their base sequences in common.

8. (a) To answer this question we must note the convention used in stating polynucleotide sequences--that is, the 5' end is written at the left. Thus, the complementary sequence to AGTC (which is really 5'-AGTC-3') is GACT (5'-GACT-3'). Hence, the frequency of CT is the same as that of its complement AG, which is given as 0.15. Similarly, AC = 0.03, TC = 0.08, and AA = 0.10. (b) If DNA had a parallel structure, both ends of each strand would be the same (say, 5'), so the sequence complementary to AGTC would be TCAG. Thus, if AG = 0.15, TC will also be 0.15. Similarly, CA = 0.03, CT = 0.08, and AA = 0.10.

9. Primer: a nucleotide bound to DNA and having a 3'-OH group. Template: a polynucleotide strand whose base sequence can be copied.

10. (a) T. (b) F. It is DNA polymerase III (c) F. A primer is also required. (d) F. There would be no template and no primer. (e) T, though *in vitro* a DNA fragment can be used.

11. As a simple repeating polymer consisting of a single unit, it cannot carry information for an amino acid sequence.

12. Use a genetic marker that is not involved in any way with polysaccharide synthesis. This objection was in fact eliminated by showing that the antibiotic-resistance phenotype could be transformed, as well as other phenotypes.

13. Remember that the group at the growing end of the leading strand is always a 3'-OH group because DNA polymerases can only add nucleotides to such a group. Because DNA is antiparallel, the opposite end of the leading strand is a 5'-P group. (a) 3'-OH. (b) 3'-OH. (c) 5'-P.

14. Rolling circle replication begins with a break that produces a 3'-OH group and a 5'-P group. The polymerase extends the 3'-OH terminus and displaces the 5'-P end. Since the DNA strands are antiparallel, the strand complementary to the displaced strand must be terminated by a 3'-OH group.

15. (a) The first DNA base copied is T, so the first RNA base is A. RNA groups by addition to the 3'-OH group, so the 5'-P end remains free. Therefore, the sequence is 5'AGUCAU3'. (b)

U; a triphosphate. (c) Right to left.

16. In conservative replication the parental strands remain together and the daughter molecule consists of two newly synthesized strands. Thus, in a density-shift experiment no hybrid DNA will ever form. After one generation there will be two bands, $^{15}N^{15}N$ (parental) and $^{14}N^{14}N$ (daughter molecule), in equal amounts.
After two generations the same two bands will be present, but the ratio of the amounts is 1:3.

17. One generation after the transfer all of the DNA is hybrid ($^{14}N^{15}N$). Continued growth for one generation yields two bands, $^{14}N^{14}N$ and $^{15}N^{14}N$, in equal amounts. After two generations the $^{14}N^{14}N$ DNA yields two identical $^{14}N^{14}N$ molecules and the $^{15}N^{14}N$ molecules produce one $^{14}N^{15}N$ molecules and one $^{14}N^{14}N$ molecules. Thus, there are two bands (light and hybrid) in a ratio of 3:1.

18. The lack of breakage has no effect after a full generation, so there are equal amounts of $^{14}N^{14}N$ and $^{14}N^{15}N$.

19. The 5'-terminal bases are obtained from the 3'-terminal bases of the complementary strand. The sequence is 5'-GATACGACGAAGTACTGG.

20. Yes, unless the RNA molecule is copied from both ends, copying only one strand from each end.

21. (a) The gap would be filled by a single piece, since there is no reason why replication could not continue from the 3'-OH group along to the 5'-P group. (b) No, for the same reason as in (a). If a 3'-OH group were not present, one could imagine a situation in which a primer was made leading to initiation within the gap. Then, the gap could be enlarged at the 3' end by a nuclease that would produce a 3'-OH group, and a second initiation event would occur by extension from this group. Alternately, the nuclease activity could come first, producing a 3'-OH group, and then there would be no fragments.

22. Because of the antiparallel nature of the individual strands of DNA, in a bidirectionally replicating molecule, the precursor fragments in the two forks will be copied from different polynucleotide strands.

23. (a) There is no maximum, because there is no stop signal for rolling circle replication; the branch grows indefinitely.
(b) A parental single strand; in particular, the one that receives a single-strand break that produces a 3'-OH group.

24. (a) The solution containing 10^{-7} mg/ml DNA would contain 0.02 percent of 10^{-7} mg = 2×10^{-14} g protein. The molecular weight of a protein containing 300 amino acids is 300 times the "average" weight of an amino

acid, or approximately $300 \times 100 = 3 \times 10^4$. Since the ratio (X molecules/2×10^{-14} g) equals the ratio of Avogadro's number to 3×10^4 g/mole, the number of molecules is 4×10^5 molecules. (b) The numbers do not exclude the possibility that the transformation is protein-mediated, because the maximum estimate of the number of protein molecules exceeds the number of transformants by a factor of 400.

25. There are no positive ions to neutralize the negative charges of the phosphates. The strong electrostatic repulsion causes the strands to come apart. This is a common effect in proteins also, because the repulsive electrostatic forces is greater than the attraction interactions caused by hydrogen bonding, van der Waals forces, and hydrophobic forces.

26. (a) They are joined by the normal action of polymerase I and DNA ligase, both of which remain active at 42 C. (b) No. Possibly, a few fragments being extended by polymerase I might have a few [^{3}H]thymidines very near the 3' terminus.

CHAPTER 5

The Molecular Organization of Chromosomes

CHAPTER SUMMARY

The DNA content of organisms varies widely. Small viruses exist whose DNA contains only a few thousand nucleotides, and among the higher animals and plants the DNA content can be as much as 150 billion nucleotides. Whereas it is generally true that DNA content increases with the complexity of the organism, within particular orders and genera the DNA content varies as much as tenfold. In this case, genome size is totally unrelated to complexity of the organism.

DNA molecules come in a variety of forms. Except for a few of the smallest viruses, whose DNA is single-stranded, and for some viruses using RNA as the genetic material, the DNA of all organisms is double-stranded. Bacterial DNA is circular, as the DNA of many animal viruses and of some bacteriophages. The DNA of higher organisms is always linear. Circular DNA molecules are invariably supercoiled: they are partially unwound, and the topology of unwinding along with the requirement of DNA to have ten base pairs per turn of the helix causes the circles to be twisted. Naturally occurring molecules are always negatively supercoiled with the same degree of twisting, about 1 twist per 200 nucleotide pairs. The bacterial chromosome consists of independently supercoiled domains: the independence is probably the result of proteins that bind to the DNA in a way that prevents rotation of the helix on one side of the binding site from affecting the DNA on the other side of the binding site.

The DNA of both prokaryotic and eukaryotic cells and of viruses is never in a fully extended state but is folded in an intricate way, thereby reducing its effective volume. In viruses the DNA is tightly folded, but without bound protein molecules. In bacteria the DNA is folded to form a multiply-looped structure, called a nucleoid, which includes several uncharacterized proteins that are apparently essential for folding. In eukaryotes the DNA is compacted into chromosomes, which contain several proteins and which are thick enough to be visible by light microscopy during the mitotic phase of the cell cycle. The DNA-protein complex of eukaryotic chromosomes is known as chromatin. The protein component of chromatin consists of five distinct proteins: the histones H1, H2A, H2B, H3, and H4. The latter four types aggregate to form an octameric protein containing two copies of each histone molecule. DNA is wrapped 1-3/4 turns around the histone octamer, forming a particle-like structure, a nucleosome. This wrapping is the first level of compaction of the DNA in chromosomes. Each nucleosome unit contains about 200 nucleotide pairs of which about 145 are in contact with the protein. The remaining 55 nucleotide pairs link adjacent nucleosomes. Histone H1 binds to the linker segment and draws the nucleosomes nearer to one another. The DNA in its nucleosome form is further compacted to a helical fiber, the 30-nm fiber. In forming a visible chromosome this unit undergoes several additional levels of folding, producing a highly compact visible chromosome. The result is that a eukaryotic DNA molecules whose length and width are about 50,000 and 0.002 micrometers respectively is folded in many ways to form a chromosome with a length of tens of micrometers and a width of about 0.5 micrometer.

During replication of chromatin the preexisting histone octamers remain associated with one of the parental strands, and octamers of newly synthesized histone are assembled along the other branch of the replication fork after replication has occurred.

Polytene chromosomes are found in certain organs in insects. These gigantic chromosomes consists of about 1000 molecules of partially folded chromatin aligned side by side. By microscopy they appear to consist of thousands of dark bands, each of which correspond to a single gene. Polytene chromosomes do not replicate further, and cells containing them do not divide. They are useful to geneticists primarily as morphological markers for particular genes.

Whereas DNA molecules are stable, any reagent or agent that breaks hydrogen bonds causes the two strands to separate, a process called denaturation. Since denaturation is often carried out by heating a DNA sample, the term melting is also used, and a plot of the degree of dissociation (fraction of base pairs that are broken) versus temperature is a melting curve. The temperature at which half of the base pairs are disrupted is called the melting temperature; its value increases with the G+C-content of the DNA and can be used to measure G+C-content. Fully separated complementary strands can rejoin and form double-stranded DNA, a process called renaturation. The rate of renaturation increases as the concentration of the DNA increases. Analysis of the association rate, known as Cot analysis, gives information about the number of copies of individual base sequences in a DNA molecule. In prokaryotic DNA most sequences are unique. However, in eukaryotic DNA, only a fraction of the DNA consists of unique sequences. Many sequences are present in enormous numbers, hundreds to millions of copies. The highly repetitive sequences are primarily located in the centromeric regions of the chromosomes and in the telomeres, the termini of the chromosomes. The significance of the repetition is not known. A large fraction of the DNA, the middle repetitive DNA, consists of sequences of which a few hundred copies per cell are present. Much of middle repetitive DNA in the higher eukaryotes consists of transposable elements, sequences able to move from one part of the genome to another. A typical transposable element is a sequence of several thousand nucleotide pairs terminated by two short sequences (10 to 300 nucleotide pairs, depending on the particular transposable element). These sequences either are identical (direct repeats) or one is inverted compared to the other (inverted repeat). These termini plus an internal gene specifying an enzyme are necessary for the movement of these elements, a process known as transposition. The first transposable elements discovered were the *Ds* and *Ac* elements in maize. Since then hundreds of such elements have been discovered in *Drosophila,* yeast, other eukaryotes, and bacteria.

Mitochondria and chloroplasts, which are cytoplasmic organelles, also contain DNA molecules. These molecules are independent of the nuclear chromosomes and carry genes for structure and function of the particle. These particles are transmitted from one generation to the next by an apparently random process, rather than the highly organized mechanism of mitosis. Occasionally daughter cells fail to receive the organelles. Segregation of these organelles does not obey the rules of Mendelian segregation and is referred to as cytoplasmic inheritance. The DNA of most organelles is double-stranded and circular. Some histone is present in these organelles, but chromatin does not form, and the structure of the chromosomes resembles the bacteria chromosome more than that of eukaryotes.

BOLD TERMS

Ac, Activator, chloroplasts, chromatin, chromatosome, chromocenter, condensed state, conserved, copia, core particle, Cot, Cot curve, covalent circle, cytological hybridization, cytological map, denatured, direct repeats, *Dissociation, Ds,* euchromatin, families, folded chromosome, genome, heterochromatin, histones, H1, H2A, H2B, H3, H4, *in situ,* inverted repeats, main band, maternal inheritance, melting curve, melting temperature, mitochondria, molecular hybridization, negatively supercoiled, nick, nonhistone chromosomal proteins, nucleoid, nucleosomes, organelles, polytene chromosomes, protamines, reannealing, relaxed, renaturation, repetitive sequences, satellite bands, scaffold, single-copy sequences, supercoiled, supercoiling, superhelicity, target sequence, telomeres, 30-nm fiber, transposable elements, transposase, transposition, unique sequence.

ADDITIONAL INFORMATION

Little more can be said about the structure of chromosomes other than the outline given in the text. The structure of nucleosomes is quite well known, as is the structure of the 30-nm fiber. Beyond that is a great deal of speculation. What is not yet clear is whether the compaction of chromatin following that of the 30-nm fiber, is a feature of chromatin alone--for example, some interaction of nucleosomes--or whether other proteins or small molecules participate in the folding. Small amounts of RNA are found in isolated chromosomes; whether these molecules are involved in chromosome structure or are simply transcripts trapped in the chromosome matrix is not known. Other proteins, known generally as nonhistone proteins, are also present in chromosomes. Some of these are certainly involved in gene expression, and others may aid in folding. Chromosome structure is a subject of active research, and presumably by the time of the second edition of the text more information will be available. These least-understood feature of chromosome structure is the mechanism or signals by which chromosomes change structure in various phases of the cell

CHAPTER 5

cycle. Reference to the micrographs in Chapter 2, which shows cells in various stages of meiosis, show that the degree of condensation changes markedly. Not shown are the interphase nuclei, in which the chromatin is about as extended as possible within the nucleus.

Denaturation was introduced in this chapter as a means of preparing DNA to be renatured. However, study of denaturation has contributed in a significant way to our understanding of DNA structure and is in fact the major source of information about the forces that stabilize the double-stranded structure. Figure 5-17 of the text showed a single melting curve that demonstrated that the optical absorbance of a DNA solution increases with temperature and that accompanying this increase is breakage of hydrogen bonds and general disruption of the molecule. Ultimately, in the uppermost portion of the curve (in the plateau region), the two strands separate. It should be noted that as hydrogen bonds break and the structure becomes more open, the double helix must unwind. The drawings in the figure oversimplify the process by showing opening without unwinding. The unwinding of a molecule with thousands of turns of a helix was once thought to be the limiting factor in the rate of DNA replication. In fact, in the late 1950s, a calculation was performed (now known to be incorrect) that showed that it would take years to unwind an *E. coli* DNA molecule. Since it only takes *E. coli* 40 minutes to replicate, the conclusion was that the DNA could not be a double helix of the type Watson and Crick proposed. A variety of experimental techniques have yielded actual numbers for the rate of unwinding, roughly 25,000 base pairs per second in the range of temperatures found in biological systems. Since the replication rate is about 2000 base pairs per second, the rate of unwinding is clearly not a limiting factor in DNA replication. In the discussion of renaturation it was pointed out that renaturation is a slow and concentration-dependent process. The time for renaturation is not the rewinding time but the time required for two complementary molecules to find one another; the zipping-up process is very rapid compared to this time. It should be clear that denaturation is independent of concentration, and since the molecules whose denaturation is being studied using have little more than 50,000 base pairs, denaturation can be considered to be, for all practical purposes, instantaneous.

It was just said that study of denaturation has yielded information about DNA structure. A few examples will now be given. An early result was that the melting temperature of DNA depended on GC content; with increasing GC content, the melting temperature increases. The interpretation of this simple fact was that hydrogen bonds were at least in part responsible for holding the strands together. A GC pair has three hydrogen bonds and an AT pair has two; therefore, a GC pair should be more stable than an AT pair, and this stability is reflected in the higher temperature required to disrupt the base pairs. Another result was that the melting temperature decreases with the concentration of the ions in the solution. Since ions tend to neutralize charge, it was obvious that charge repulsion of some kind is important in the DNA structure. In studies of pairing of free bases or nucleosides no affect of salt concentration was observed, which indicated that the charge did not reside in either the base or the sugar. This result made it obvious that the powerful negative charges of the phosphates of the backbone repel one another, so a stable DNA molecule can exist only when these charges are neutralized. Studies of mixtures of DNA and positively charged protein molecules (for example, histone molecules) showed that the charge could be neutralized either by positive ions or by the protein molecule. Addition of reagents that eliminate hydrogen bonds in solution, for example, formamide and urea, cause a substantial decrease in the melting temperature; these results gave further support to a role of hydrogen bonds in the DNA structure. An initially surprising result was that the addition of reagents that destabilize hydrophobic interactions (the clustering of uncharged molecules in water) also causes the melting temperature to decrease, from which it was necessary to conclude that hydrophobic interactions also contribute to DNA structure. A clue to understanding this phenomenon came from a study of the effect on the structure of single-stranded polynucleotides of reagents that disrupt hydrophobic interactions; these reagents increase the flexibility of the single strand and also eliminate optical activity. Optical activity was shown to be a result of stacking of the bases, an array in which the planes of the bases are parallel and rotate slightly following a helical path. These experiments indicated that base stacking occur in double-stranded DNA. Examination of the entire set of data concerning hydrogen bonding and hydrophobic interactions yielded the following understanding of the forces that stabilize the double-stranded structure. The ring structure of the DNA bases makes the bases only weakly soluble in water, and the low solubility causes the molecules to form clusters rather than interact with water. This tendency to cluster is a hydrophobic interaction and is the cause of the parallel stacking of the bases. Hydrogen bonding is directly responsible for the pairing of bases and hence for the pairing of the individual strands. However, hydrogen bonds are weak forces and require correct orientation. The stacking of the bases causes an orientation of the hydrogen-bonding groups that enable many hydrogen bonds to form. Similarly, the hydrogen bonds cause an orientation of the bases that stabilizes base stacking. Thus, base stacking is maximal when there is hydrogen bonding and hydrogen is maximal when there is base stacking, so the two interactions act synergistically to produce a stable double-stranded structure. The stacking primarily determines the shape of the single strands, and the hydrogen-bonding is responsible for alignment and interaction of the strands.

DRILL QUESTIONS

1. How many individual protein molecules are in a nucleosome?

2. Which type of histone does not exist in nucleosomes in pairs?

3. Are all repetitive sequences transposable elements?

4. If you were determining the base sequence of a long DNA sequence, what features of the sequence would make you confident that a transposable element was present?

5. If you had never seen a chromosome with a microscope but had seen cells with nuclei, what information would nonetheless let you know that DNA must exist in a highly coiled form within cells?

6. Are the ratios of the different histones to one another the same in all cells of a eukaryotic organism? In all eukaryotic organisms?

7. Is the histone:DNA ratio the same for all cells of a eukaryotic organism? In all eukaryotic organisms?

8. Are histones found in all cells?

9. What would happen to a chromosome if it were put in concentrated NaCl?

10. What is the relation between the 110-A fiber and the 30-nm fiber in chromatin?

11. What causes the bands in a polytene chromosome?

12. What is the relation between a band in a polytene chromosome and a gene?

13. How do polytene chromosomes replicate?

14. Distinguish denaturation and renaturation with respect to DNA and to macromolecules in general. For DNA, which process is concentration-dependent?

15. Two DNA molecules have molecular weights of 5 and 10 million. If samples of each at a concentration of 20 micrograms per milliliter are separately denatured and renatured, which one will denature more rapidly?

16. A particular Cot curve has two steps. What information does this fact provide?

17. What kind of genes are found in heterochromatin?

18. What is the location of a telomere on a chromosome?

19. Is there an analogue of a telomere in the *E. coli* chromosome?

20. What is the basic sequence organization of a transposable element? Mention direct and inverted repeats.

21. What types of genes in eukaryotes are never inherited according to the rules of Mendelian genetics?

ANSWERS TO DRILL QUESTIONS

1. There are four distinct molecules, each of which exists in two copies, forming the histone octamer, plus a single protein in the linker.

2. H1.

3. No, for example, the repeated sequences in telomeres and centromeres.

4. A unique sequence flanked by a sequence in direct or inverted repeated (the termini of the transposable element), the entire three-component unit flanked by a short sequence in direct repeat (the target sequence).

5. From the total molecular weight you would know the total length of the DNA in a nucleus. Even allowing for up to 100 distinct DNA molecules per cell, the length of each molecule would still be very great compared to the diameter of a nucleus. You would be forced to conclude that each DNA molecule must fold back on itself repeatedly.

6. There is always a 1:1:1:1 ratio of histones H2A, H2B, H3, and H4 in all eukaryotic cells because of the universality of the histone octamer. The ratio of H1 to the octamer varies slightly.

7. No, because the size of the linker varies from one cell type and from one organism to the next.

8. No, not in prokaryotes and not in sperm. Sperm have protamines instead.

9. The histones would no longer be able to bind to the DNA, so the entire chromosome would be disrupted and the DNA freed.

10. If a 110-A fiber is arranged in helical form with about six nucleosomes per turn, the result is a 30-nm fiber.

11. Matching of chromomeres, which are locally folded regions of chromatin.

12. Each band represents one gene.

13. They do not replicate.

14. In general, denaturation refers to loss of three-dimensional structure and renaturation refers to restoration of the structure destroyed by denaturation. For DNA, denaturation refers to breakdown of the double helix and ultimate separation of the individual strands. Renaturation is the formation of double-stranded DNA from complementary single

CHAPTER 5   75

strands. Renaturation is concentration-dependent.

15. The smaller one.

16. There are at least two frequency classes, and there is at least one sequence that is repeated.

17. Generally, genes are not present in heterochromatin.

18. Telomeres are at the termini.

19. No, because the *E. coli* chromosome is circular and hence has no termini.

20. A sequence of several thousand nucleotide pairs flanked by a copy of a shorter sequence. If the flanking sequences have the same orientation, they are said to be in direct repeat; if their orientations are reversed, they are said to be in inverted repeat.

21. Those genes carried on the chromosomes of organelles.

ADDITIONAL PROBLEMS

1. Two morphologically distinct classes of chromatin are found in most cells, heterochromatin and euchromatin. Studies of the replication of chromatin have indicated that heterochromatin replicates very late in the replication cycle of a cell compared to euchromatin. This late replication might be understandable if the heterochromatic segments were always at the ends of the linear DNA molecule of a eukaryotic chromosome, but this is not the case. Suggest some mechanism or some feature of the heterochromatic DNA sequences that might cause delayed replication.

2. Why do you think that heat causes denaturation of DNA?

3. DNA is isolated from a bacterial culture, but during the isolation the molecule are broken into fragments about one percent of the size of the intact DNA molecule. The DNA is denatured and renatured and then treated with an enzyme that can cleave only single-stranded DNA. Would any DNA be degraded?

4. Would the Cot curve of the DNA of haploid yeast differ in any way from that for the DNA of diploid yeast?

5. The following experiment is carried out to determine the extent to which two bacterial species have common DNA sequences. The DNA of bacteria I is isolated (and, as usual, extensively fragmented during the isolation), denatured, and made to adhere to a nitrocellulose filter. Radioactive DNA from bacteria II is also isolated and denatured. The radioactive DNA and the nitrocellulose filter bearing the nonradioactive DNA of bacteria I are mixed and subjected to conditions that allow renaturation to occur. If no DNA of

bacteria I has previously been attached to the filter, no radioactive denatured DNA will stick to the filter. If a small amount of denatured radioactive DNA of bacteria I is added to a filter containing nonradioactive denatured DNA of bacteria I, all of the radioactivity will bind to the filter (in other words, renaturation occurs perfectly well on a filter). After incubation of the filter with the denatured radioactive DNA of bacteria I and washing of the filter to remove unrenatured radioactive DNA, it is found that 22 percent of the radioactivity remains on the filter. In a separate experiment the renatured filters are treated with an endonuclease that degrades only single-stranded DNA and then washed to remove unbound radioactivity; in this case only 9 percent of the radioactivity is bound to the filter. What was the purpose of the nuclease treatment, and what fraction of the DNA of bacterium II is homologous to the DNA of bacterium I?

6. A haploid yeast species has two mating types 1 and 2. Mating occurs by fusion of the haploid cells to form a stable diploid. The diploid can be induced to form four ascospores in an ascus. A true-breeding haploid mutant is isolated that produces colonies with an unusual appearance. Mutant and wildtype are mated. The diploid colonies are normal-looking. When ascospores are produced and the individual spores are removed from the ascus and allowed to form colonies, almost all asci contain four spores that produce normal colonies. What is the probable mode of inheritance of the mutant trait, assuming that it is caused by a single gene?

ANSWERS TO ADDITIONAL PROBLEMS

1. Recall from Chapter 4 of the text that eukaryotic DNA is replicated from a very large number of origins. Each origin must be recognized by an initiator substance, presumably a protein. If the replication origins of heterochromatic DNA were different from those of euchromatic DNA, the timing of initiation could easily be different. However, whether all replication origins in the DNA of a particular eukaryotic chromosome are identical or of two types is not known, nor is there any evidence on the point. What seems likely though is that the structure of heterochromatin is sufficiently different from that of euchromatin that the DNA is simply not as easily available to the initiator molecule. Possibly, the DNA is bound either to histones plus other proteins or just to other proteins in the heterochromatic region and that this binding is so tight as to reduce access to the initiator. Either a structural change or synthesis late in the cell cycle of a more powerful initiation protein could account for the difference in timing.

2. Hydrogen bonds are fairly weak. Heating causes molecular vibration and movement, and the kinetic energy produced is great enough to move the bases sufficient far apart that hydrogen bonding cannot be maintained. Once one base pair is disrupted, adjacent ones are more easily disrupted.

3. Yes. Since the DNA has been fragmented at random, complementary sequences will not always be located in fragments of the same size. Therefore, a sequence in one molecule can anneal with the complementary sequence in a longer molecule leaving terminal single-stranded regions. These would be attacked by the nuclease.

4. For equal total concentrations of DNA the diploid curve would be shifted to the left, because there would be twice as many copies of all sequences as in haploid DNA. The shapes of the two curves would be identical though.

5. As pointed out in Additional Problem 3, the base-paired regions may be only a fraction of each renatured fragment, so the amount of radioactivity bound would overestimate the fraction of DNA that is homologous. Treatment with the nuclease removes the unpaired single-stranded segments. Thus, 9 percent of the DNA of bacterium II is homologous to that of bacterium I.

6. Inheritance is definitely non-Mendelian. It is not surprising that the diploid, which is heterozygous should produce normal colonies, if the mutation is a recessive. However, one would expect 2:2 segregation of normal and unusual colonies among the four spores of an ascus. Since segregation is 4:0, all cells contain a wildtype gene. The most likely explanation is that the trait is carried by an extrachromosomal cytoplasmic element that exists in fairly high copy number in each cell. The cytoplasm of the wildtype is present in the diploid. In a subsequent the element is segregated with the cytoplasm, about equally to each haploid cell.

SOLUTIONS TO PROBLEMS IN TEXT

1. (a) There are 10 base pairs for every turn of the double helix. With four turns of unwinding, $4 \times 10 = 40$ base pairs are broken. (b) Each node represents one full turn of the helix; hence, four nodes. (c) A supercoil with one node--that is, a figure-8. (d) Again, 10 base pairs per turn means 50/10 or five turns of unwinding and hence five turns of supercoiling. Since the molecule is already in the underwound state and the underwinding is now stabilized, the supercoiling is in the opposite sense from the normally negative supercoiling of the DNA that existed before the pairs per broken. Hence the five turns are positive.

2. The supercoiled form is in equilibrium with an underwound form having many unpaired bases. At the instant that a segment is single-stranded, S1 can attack it. Since a nicked circle lacks the strain of supercoiling, there is no tendency to have single-stranded regions, so a nicked circle is resistant to S1.

3. The A+T-content is all that is important. The lowest A+T-content has the highest melting temperature, so the order is 3, 2, 1.

4. In contrast with problem 3, the length of the G+C tracts are important. Molecule 2 has a long G+C segment which will be the last region to separate. Hence molecule 1 has the lower temperature for strand separation.

5. Remember that reversibility will be complete until the last hydrogen bond is broken, that is, until the strands separate completely.

Figure 1

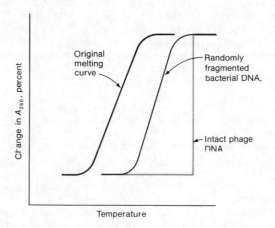

6. Recall from the 1:2:1 ratio observed in random assortment that if two collections of element can both self-combine and cross-combine, only half are hybrid. Here, five percent is hybrid, representing 10 percent (twice five) of the sequences.

7. (a) The DNA is about 30 percent unique, about 60 percent redundant, and the remaining 10 percent is highly redundant, possibly satellite, DNA. (b) The three classes have Cot midpoints of about 10^3, 10^{-1}, and 10^{-3}, respectively. Thus, the number of copies per genome is $10^3/10^3 = 1$ for the unique DNA (by assumption), $10^3/10^{-1} = 10^4$ for the redundant DNA, and $10^3/10^{-3} = 10^6$ for the highly redundant DNA.

8. Chromatin--the DNA-histone complex present in eukaryotic chromosomes. Histones--a class of five proteins bound to DNA in chromatin. Nucleosome--A DNA-histone complex consisting of a DNA-histone core particle, linker DNA, and histone H1. Core particle--a DNA-histone complex consisting of an octamer (two

copies each of H2A, H2B, H3, and H4) and a 140-nucleotide-pair DNA molecule. Chromatosome--a core particle plus histone H1, in which the DNA is wrapped around the histone octamer in two complete turns.

9. The fact that binding is weak in 1 M NaCl suggests an electrical attraction is involved in the binding. The charges on the bases are sufficiently weak (and mostly engaged in hydrogen bonding) that only the highly charged phosphate groups of the sugar-phosphate backbone seem to be candidates. Weak binding to single-stranded DNA is consistent with this conclusion, because a randomly coiled single strand would not present an ordered array of charges to the protein; the rigidity of double-stranded DNA provides such order. Therefore, the protein undoubtedly binds to the phosphates.

10. In the first experiment, nucleosomes have formed at random. In the second experiment, P has bound to a unique sequence and nucleosomes have formed by sequential addition of histone octamers from the site of P. Thus, the linkers must be at fixed distances from the unique binding sequence. Since nuclease attack occurs only in the linker regions, the cuts are made in small regions that are highly localized

11. Direct repeat: a sequence is repeated, each with the same 5'-to-3' orientation, that is, ABCD...ABCD. Indirect repeat: a double-stranded sequence is repeated in the reverse orientation, and because of the antiparallel nature of DNA, the sequence is ABCD...D'C'B'A', in which X' is the base complementary to X.

12. The fact that the target sequence and the transposable element are duplicated during transposition. Recently, it has been suggested that the type of replication in eukaryotes differs from that in prokaryotes. It seems clear that a form of normal DNA replication occurs in prokaryotes (see also Chapter 8). However, in eukaryotes it has been proposed, though data are quite meager, that (1) an RNA copy of the transposon might be made and released, (2) the RNA serves as a template for synthesis of a complementary single strand of DNA, which is then converted to double-stranded DNA, and (3) the double strand is then inserted, completing the transposition process.

13. Remember that the base composition determines the value of T_m but not the temperature at which strands come apart. The latter is determined by the thermal stability of the most stable tract of bases. A small number of GC pairs in a row provide greater stability toward strand separation than a larger number of GC pairs that are scattered throughout the DNA. Thus, the GC pairs in molecule I are probably dispersed; in contrast, those of molecule II are clustered, so that long segments of stable GC regions exist.

14. This situation differs from that of problem 13. As stated there, the value of T_m is affected by base composition, not by base sequence. Therefore, both species will have the same melting temperature.

15. One single-strand break (a nick) causes loss of supercoiling. Thus, if it is a dimer, in time supercoils will be replaced by open circles, without any intermediates appearing. If it consists of linked circles, each will be independent; hence, one single-strand break will convert the structure to a supercoil linked to an open circle.

16. Many copies of chloroplasts are present in a single green plant cell. Thus, Cot analysis should show a major fraction of repetitive DNA. Since after period of growth in the dark chlorophyll is lost but the Cot curve is unchanged, chloroplast DNA must still be present. Therefore, chloroplasts and their DNA remain, even though the chlorophyll is lost.

17. The DNases probably bind to the negatively charged phosphates, which are neutralized by the Na^+ ions in 1 M NaCl.

18. The amino acid sequence of histones is nearly the same in all organisms along the evolutionary scale, suggesting that functional histone is extremely intolerant of amino acid changes and that all changes are probably lethal.

19. The relatively large number of bands and the relocation in one fly suggests that the sequence may be a transposable element.

CHAPTER 6

Variation in Chromosome Number and Structure

CHAPTER SUMMARY

The chromosomes of most organisms contain a single discrete centromere, the position of which determines the shape of the chromosome as it is pulled to the poles of the cell during anaphase. Chromosomes with no centromere, or those with two or more centromeres, are usually lost within a few cell generations because of aberrant separation during anaphase.

Some organisms normally contain more than two complete sets of chromosomes; these are called polyploid. Polyploidy is widespread among higher plants and uncommon otherwise. An autopolyploid organism contains multiple sets of chromosomes from a single ancestral species; allopolyploid organisms contain complete sets of chromosomes from two or more ancestral species.

Organisms occasionally arise in which an individual chromosome is either missing or present in excess; in either case, the number of copies of genes in such a chromosome is incorrect. Departures from normal gene dosage often result in reduced viability of the zygote in animals or of the gametophyte in plants.

The normal chromosome complement in humans consists of 22 pairs of autosomes, which are assigned numbers 1-22 from longest to shortest, and one pair of sex chromosomes (XX in females and XY in males). A zygote containing an abnormal number of autosomes usually fails to complete normal embryonic development, though zygotes with Down syndrome (trisomy 21) often live for several decades. Individuals with excess sex chromosomes survive, because the Y chromosome contains few genes and because only one X chromosome is genetically active in cells of females.

Most chromosomes with abnormal structures are of one of four types--duplications, deletions, inversions, and translocations. Chromosomes with a duplication have two copies of a chromosomal segment containing one or more genes or a portion of a gene. A chromosome with a deletion is missing one or more genes. Imbalance of gene dosage resulting from small duplications or deletions can often be tolerated by the organism, but large duplications or deletions are almost always harmful. A chromosome with a group of adjacent genes in reverse of the normal order is said to contain an inversion. Since expression of the genes is usually unaltered, inversions rarely affect viability. However, pairing and crossing over between an inverted chromosome and its noninverted homologue during meiosis results in the formation of abnormal chromatids containing a duplication of some genes and a deletion of others. If the inverted region does not include the centromere, then the abnormal chromatids will contain either two centromeres or no centromere.

Two nonhomologous chromosomes that have undergone an exchange of parts constitute a reciprocal translocation. Organisms that contain a reciprocal translocation as well as the normal homologous chromosomes of the translocation produce fewer offspring (this is called semisterility) because of abnormal segregation of the chromosomes during meiosis. This semisterility can be mapped genetically, like any other genetic locus, and its map position represents the breakpoint of the reciprocal translocation. A translocation may also be nonreciprocal. A Robertsonian translocation is a type of nonreciprocal translocation in which the long arms of two acrocentric chromosomes are fused.

Malignant cells in many types of cancer contain specific types of chromosome abnormalities. Frequently, the breakpoints coincide with the chromosomal location of one of a group of genes (oncogenes) whose products have been implicated in cancer.

Although most genes function normally without being affected by their position either in the chromosome complement or in a particular chromosome, position effects in gene expression in *Drosophila* can occur when a gene that is normally present in euchromatin is moved to a position in or near heterochromatin.

BOLD TERMS

acentric, acrocentric, adjacent-1 segregation, adjacent-2 segregation, allopolyploid, alternate segregation, amniocentesis, aneuploid, autopolyploidy, Barr body, colchicine, deletion, dicentric, dosage compensation, double-Y syndrome, Down syndrome, duplication, euploid, hexaploid, inversion, karyotype, Klinefelter syndrome, knob, metacentric, monosomic, multivalent, octaploid, oncogene, paracentric inversion, pericentric inversion, Philadelphia chromosome, polyploidy, polysomy, position effect, reciprocal translocation, retroviruses, Robertsonian translocation, submetacentric, tandem duplication, tetraploid, triplication, triploid, trisomic, trisomy-X syndrome, Turner syndrome, unequal crossing over, V-type position effect.

ADDITIONAL INFORMATION

First, a few words need to be said about the terminology used to describe chromosomes--that is, acentric, acrocentric, and so forth. These terms, which may seem intimidating and/or hard to remember, have no biological or genetic significance. They are merely convenient terms to describe the appearance of a particular chromosome, especially when distinguishing it from other chromosomes. The terms are in common use though, and should be learned.

The notation used to describe various chromosome abnormalities are also worth reviewing. The terms monosomy or trisomy followed by a number indicate the chromosome that is lacking or the extra chromosome. For example, monosomy 3 would mean that the cell is lacking one copy of chromosome 3 (or, alternatively, has one copy). Trisomy 13 means that cells contain three copies of chromosome 13. Note that the word disomy is rarely used; it would mean a normal cell. An alternative notation is also employed to describe monosomic and trisomic individuals: the number of chromosomes present in cells of the organism is stated, followed by the number of copies of the particular chromosome of interest. Thus, for a human, which normally has 23 pairs or 46 chromosome, the notation 45,X indicates that the cells have only 45 chromosomes and the missing chromosome is the X. This individual could also be designated monosomy X, or XO. Similarly, a 47,XXX individual has 47 instead of 46 chromosomes, and the extra one is an X, because XXX says that three X chromosomes are present. A female trisomic for chromosome 21 could also be designated 47,XX+21, and a male monosomic for chromosome 3 could be denoted 45,XY-3. Portions of chromosomes are sometimes absent. The p,q notation describes this anomaly. For chromosomes having a short arm and a long arm, the short arm is denoted p and the long arm q. Thus, the short arm of chromosome 8 would be written 8p. A deletion of the short arm of chromosome 8 is denoted by 8p-. Combining several of the notations just presented, we note that an XXY individual lacking the long arm of chromosome 7 would be designated 47,XXY,8q-.

In Section 6.4 of the text the Giemsa technique for staining chromosomes is mentioned. This procedure is one of several staining protocols that produces bands in chromosomes. The structural and physicochemical significance of these bands is not known, though they certainly represent variations in the ability of different regions of the chromosome to bind various molecules (the stain), probably because of differences in the mode of packing of the chromatin. Other procedures (for example, quinacrine staining) produce patterns of bands that differ from the Giemsa bands. These bands are useful primarily for identifying chromosomes, since in most organisms that have more than a few chromosomes, several different chromosomes have the same appearance by simple light microscopy. However, these chromosomes are invariably distinguishable by their unique patterns of banding. Translocations can also be identified by noting unusual patterns of banding.

An important point must be made about polytene chromosomes. These chromosomes are dead ends, as pointed out in Chapter 5, in the sense that they are not capable or further replication. They are interesting mainly as a tool for cytological confirmation of certain genetic phenomena and as a means of locating genes. Some of these uses have been documented in the chapter but a few more will be reviewed here. The polytene chromosomes of *Drosophila* (and other Dipteran flies) have so many recognizable features, such as bands, bulbs, knobs, and constrictions, that small deletions, translocations, and inversions of chromosomal segments can be detected. Study of small deletions of known genes has enabled one to match bands with genes. Extensive analyses of this sort has resulted in a matching of almost all known *Drosophila* genes with particular bands. One conclusion drawn from this cytological map is that the order of the chromosomal and genetic maps are identical; however, the genetic distance between genes is not always the same as the physical distance, which means that the frequency of recombination along the length of the chromosome is not a constant.

This chapter include a discussion of a variety of abnormalities of the number of sex chromosomes in humans.

There is no need to review that subject, but it is noteworthy (and curious) that it is possible, though exceedingly unusual, for monozygotic twins to have different sexes. This rare event occurs when a male zygote (XY) undergoes nondisjunction, forming an XXYY cell, which then divides to form a XYY fertile male cell and and XO (Turner) sterile female cell.

DRILL QUESTIONS

1. What kind of chromosome cannot move during anaphase?

2. What type of chromosome forms a bridge at anaphase?

3. What are the three types of chromosomes, defined by their shapes during anaphase?

4. What term is used to describe a cell with three sets of chromosomes?

5. What term is used to describe a cell with two sets of chromosomes but three copies of one homologue?

6. Do polyploids have normal viability in plants?

7. Name three ways a nonhaploid cell can have an odd number of chromosomes.

8. How can a gamete be diploid?

9. A hybridization of a diploid species Q with another diploid species R yields tetraploid species. What term would be used to describe the species?

10. What gametes are produced by a tetraploid parent with the genotype $AAaaBBbb$?

11. What gametes are produced by the trisomic A_1A_2a, and with what frequencies?

12. What gametes are produced by the trisomic $AAaBb$ and in what frequencies?

13. Why are monosomics highly defective and usually inviable?

14. What is the defect in Down syndrome?

15. Name two types of individuals that could have one Barr body per somatic cell.

16. What protective mechanism in humans keeps the number of individuals with major chromosomal abnormalities low?

17. Why are very large deletions usually very defective, even in a heterozygous individual?

18. In *Drosophila*, what is the relation between band number

in polytene chromosomes and gene number?

19. Name two methods, one microscopic and one genetic, to detect a deletion in Diptera (two-winged flies).

20. If a chromosome has segments ABCDEFG, what is the sequence if there is a C-E inversion and if there is a C-E deletion?

21. How can a triplication arise and what product usually accompanies its formation?

22. Consider a chromosome with the sequence ABCDXEFGHI, in which X is the centromere. Write a sequence for a pericentric inversion including two segments of the chromosome.

23. Two chromosomes with the sequences ABCDEFG and MNOPQRSTUV, respectively, undergo a reciprocal translocation at the junctions E-F and S-T. What are the products?

ANSWERS TO DRILL QUESTIONS

1. An acentric, which has no centromere and hence no site for attachment to the spindle.

2. A dicentric. Each centromere moves to opposite poles.

3. Metacentric (central centromere), submetacentric (noncentral centromere), acrocentric (nearly terminal or terminal centromere).

4. Triploid.

5. Trisomic.

6. Yes.

7. Triploidy with odd number of chromosomes in the haploid set, monosomy in a diploid, trisomy in a diploid.

8. If the parent is tetraploid.

9. Allopolyploid.

10. AABB, AABb, AAbb,
    AaBB, AaBb, Aabb,
    aaBB, aaBb, aabb

11. 1/6 *AA*, 2/6 *Aa*, 2/6 *A*, 1/6 *a*.

12. 1/12 *AAB*, 1/12 *AAb*, 2/12 *AaB*, 2/12 *Aab*,
    2/12 *AB*, 2/12 *Ab*, 1/12 *aB*, 1/12 *ab*.

13. With only one copy of a particular chromosome, every recessive allele is expressed.

14. An extra chromosome 21, as a trisomic or as a Robertsonian

translocation.

15. A normal XX female, and and XXY or XXYY male.

16. Spontaneous abortion.

17. All of the recessive alleles in the region of the undeleted chromosome that is deleted in the homologue are expressed.

18. They are equal.

19. Microscopic: look for a missing band in a polytene chromosome. Genetic: in a cross the lack of production of a wildtype phenotype with many nearby recessive alleles is highly suggestive of a deletion covering the region including those alleles.

20. Inversion: ABEDCFG. Deletion: ABFG.

21. Unequal crossing over between two duplications yields a triplication and a normal chromosome with only a single copy of the triplicated segment.

22. ABCEXDFGHI.

23. ABCDETUV and MNOPQRSFG.

ADDITIONAL PROBLEMS

1. If blue skin color were determined by an X-linked allele responsible for synthesizing a blue pigment, and yellow skin color were an allelic X-linked trait (the allele is responsible for making yellow pigment), would you ever expect to find a heterozygous female with green skin? Assume that the alleles are so near that crossing over between them can be neglected.

2. Give an argument that maleness in humans is determined by the presence of the Y chromosome rather than lack of an X.

3. Color-blindness is an X-linked trait. A color-blind 45,X woman with Turner syndrome has a color-blind father. Which parent donated the aberrant gamete that led to the Turner syndrome?

4. What is the probable genotype of a color-blind male with 47,XXY Klinefelter syndrome? Also, if both parents had normal vision, what are their genotypes?

5. The band patterns of the polytene chromosomes of five different subpopulations of a fly found in distinct Georgia swamps are being studied. A particular region is found to have a different band order in each population. The band patterns are the following (each band has a letter): 1. ecafhdbg; 2. ehdcafb; 3. ehfacdbg; 4. cafhedbg; 5. ehfbdcag. Assuming that the populations differ from one another only in gene

inversion, explain how these populations are related, indicating which population was probably the parent one.

6. Japanese and Chinese dragons are being studied. Though they have many traits in common, there are enough differences to suggest that they are different species. The genes for tail length and for breath temperature in the Japanese strain are five map units apart. In the Chinese strain they were unlinked and, in fact, on different chromosomes. How might you explain this species difference?

7. How can you tell by microscopy whether a tetraploid plant discovered in the Amazon jungle is an allopolyploid or an autopolyploid? Assume that the individual chromosomes are difficult to distinguish from one another.

8. Why is it much easier to establish true-breeding strains of tetrasomic organisms than of trisomic ones?

9. Other than trisomy-21, all human autosomal trisomies are lethal before or just after birth. Why is this not also true of the sex chromosomes?

ANSWERS TO ADDITIONAL PROBLEMS

1. No, because of random inactivation of the X chromosome. Since the two alleles are on homologues, no somatic cell can have both X chromosomes active. Thus, the organism will have patches of blue skin and patches of yellow skin.

2. An XXY individual is male.

3. The father gave her a gamete with a functional X, since he gave her his color-blindness. Therefore, it must have been her mother who produced a gamete lacking an X.

4. The male is an XXY. Since color-blindness is recessive, both X chromosomes must carry the allele for color-blindness. If the father was normal, he cold not carry the color-blindness allele. The mother has to be XX; since she is normal, she must be heterozygous for the color-blindess allele.

5. All were derived by single inversions from number 3.

6. Most likely, a translocation occurred that moved a portion of the chromosome containing only one of the genes to another chromosome.

7. Examine plant cells that are in meiosis. In meiotic metaphase the four sets of identical chromosomes in an autopolyploid often synapse as tetravalents. An allopolyploid has only two sets of chromosomes and hence can only form bivalents.

8. In a trisomic two of the chromosomes go to one pole and one

CHAPTER 6   89

to the other, and the distribution is random for each set of three. Occasionally, the unpaired chromosome will be lost. Hence, usually the zygote will suffer from chromosome imbalance. However, the four chromosomes of a tetrasomic usually synapse in twos or as a group of four, so the gametes get two members of each chromosome.

9. In humans, a mechanism exists for inactivating all X chromosomes in excess of one, thereby maintaining genic balance.

SOLUTIONS TO PROBLEMS IN TEXT

1. Telomeres are at the termini of chromosomes. If there are no telomeres, there are no termini. Hence the chromosome must be circular. Chromosomes of this type are called ring chromosomes.

2. An inversion that includes the centromere but in which the telomere is not central can move the telomere from a central location in a metacentric chromosome to a noncentral location, forming a submetacentric. A nonreciprocal translocation in which the short arm of an acrocentric fuses with the long arm of another acrocentric could produce a chromosome with two equal arms, that is, a metacentric; such a nonreciprocal exchange is a Robertsonian translocation.

3. An autopolyploid series is formed by merging of identical sets of chromosomes. Therefore the chromosome numbers must be multiples of the haploid number, or 10 (diploid), 20 (tetraploid), 30, 40, and 50.

4. The chromosome number was doubled. The zygote probably began as 46,XY. Failure of chromosome separation in the first meiotic division results in a doubling of the entire chromosome complement.

5. The semifertile one would have had one haploid set of each type, or 18 chromosomes in all. Since the rabbage is tetraploid, it must consist of two complete diploid sets, or 36 chromosomes.

6. Species A and B hybridized and the chromosome number in the hybrid was double. Therefore, S is an allotetraploid of A and B.

7. Since only one X is active in any particular cell, the X must not carry absolutely lethal mutations. Lacking any chromosome but an X would probably prevent embryological development. Thus, the fetus was probably 45,X, and had it lived, it would have had Turner syndrome.

8. Since it is extremely unlikely that the mother could have been a monosomic and lived to adulthood, it is likely that the missing chromosome is fused to another chromosomes. That is, she must have a Robertsonian translocation. A child with Down

syndrome would have to have three copies of chromosome 21. Thus, the child must have one copy from the father and one from the mother and probably the third also from the mother. The source of the third could be her Robertsonian translocation, in which one of the participants was certainly chromosome 21. Thus, the affected child would have 46 chromosomes, the extra chromosome-21 component fused to another chromosome.

9. Reduced recombination in a chromosome is invariably indicative of a large inversion. Pairing of a chromosome with an inversion with a normal homologue, yields acentrics and dicentrics in some meiotic events. Thus, the inbred strain is homozygous for a large paracentric inversion in chromosome 6.

10. The first uncovers a, b, and d, so neither c, e, nor f can be within the region defined by a, b, and d. The second uncovers a, d, c, and e, without uncovering b, so that b cannot be between a and d, nor in the $a\ d\ c\ e$ region. Thus, b is on one side of a-d and c, e, and f are on the other side of a-d. The third uncovers e and f, but not the others, so f and a-d-c are on opposite sides of e. Thus, the order is $b\ a\ d\ c\ e\ f$ or $b\ d\ a\ c\ e\ f$. None of the deletions tell anything about the order of a and d with respect to b or to c.

11. Because of aneuploid (chromosomally unbalanced) gametes resulting from adjacent-1 and adjacent-2 segregation of heterozygous translocation; translocation homozygotes form two bivalents instead of a quadrivalent; in the F_1, all will be translocation heterozygotes and therefore semisterile.

12. The species have homologous chromosomes, but they differ by a reciprocal translocation.

13. Deletion 1 shows that a, b, and c form a cluster. Deletion 2 shows that a and c are a cluster without b, so the order must be $b\ a\ c$ or $b\ c\ a$. Deletion 3 shows that a and c form a cluster without b, so $b\ a\ c$ is correct; a, c, and e are a cluster, so these four genes are in order $b\ a\ c\ e$. Deletion 4 shows that c, d, and e are a cluster, so the five genes are in order $b\ a\ c\ e\ d$. The fifth deletion shows that d, e, and f are also a cluster, so bands 1-6 correspond to genes b, a, c, e, d, f, in that order.

14. Dup6 has a normal amount of enzyme, whereas all of the others have increased amounts; band 3 is the only band that is not duplicated in Dup6 and is duplicated in all the others.

15. The normal haploid parent appears to be trisomic for chromosome 2. The wildtype strain is not really haploid but carries an additional copy of chromosome 2. Therefore, the hybrids will be trisomic for chromosome 2, and many 4:0 and 3:1 tetrads will be formed by segregation of the three chromosomes.

16. The observed recombination frequency (0.1 percent) is much less than would be expected. From the map position of the genes one would normally expect about 21,700 recombinant progeny (cn + and + cn) from this cross, instead of the 47 that were observed. Reduced recombination is almost always caused by an inversion. Thus, the female is probably heterozygous for an inversion in chromosome 2. Inversions are frequently introduced into organisms by experimenters when it is necessary to suppress crossing over.

17. The second chromosome carrying Cy underwent a reciprocal translocation with the Y chromosome. This could result from a translocation of a portion of the second chromosome, carrying the Curly locus, to the Y chromosome in the irradiated male.

18. If the wildtype y allele were carried on the single X of the wildtype male, all progeny should be heterozygous with the wildtype allele on the X. In the testcross, all females should be wildtype and all males should be mutant (having only the X of the female carrying the mutant allele). Since wildtype is instead associated with maleness, the wildtype allele cannot be carried on the X of the y^+ son. The most likely explanation is that the wildtype allele is associated instead with the Y; that is, in a sperm of the irradiated male the tip of the X carrying the wildtype allele was broken off and attached to the Y. Chromosome breakage and rejoining is a common consequence of x irradiation.

19. A two-strand double crossover occurred in the inversion loop, with one crossover between b and c and the other between d and c. Three-strand double crossovers involving the appropriate chromatids could also explain the result. The rare offspring do carry the inversion.

20. The female is homozygous recessive. The point of this problem is that the only male gametes that can give produce male progeny are those that contain the reciprocal translocation. Otherwise, some segments of the X would be absent in progeny, and such offspring would be lethal. Thus, the frequencies of the genotypes of the progeny are 1/2 wildtype females and 1/2 ruby dumpy males. All other classes of progeny will be lethal.

21. (a) The semisterile F_1 individuals are heterozygous for the translocation and for virescence ($T.+/vir$) and when crossed with a fertile virescent type (a homozygous recessive) will yield semisterile nonvirescent and fertile virescent parental types and semisterile virescent and fertile nonvirescent recombinants (from a reciprocal exchange between T and vir. Thus, the recombinant frequency is (39 + 31)/588 = 11.9 percent. (b) If the gene had been on a different chromosome, the F_1 individuals would have one chromosome with the translocation, one normal homologue, one

with the virescent allele, and one without the allele. Since these chromosomes would assort independent, four types of gametes are produced equally. Thus, the phenotypic ratios would have been 25 percent for each class of progeny.

22. Brachytic--fine stripe, $(17 + 6 + 1 + 8)/682 = 4.7$ percent; brachytic--breakpoint, $(19 + 8 + 17 + 25)/682 = 10.1$ percent; fine stripe--breakpoint $= (19 + 6 + 1 + 25)/682 = 7.5$ percent. The genetic map of chromosome 1 has the order brachytic--fine stripe--breakpoint.

23. d, a, b.

24. A B C D E; A B c; a d; a d c b e.

25. (a) There are three possibilities: (1) There is a paracentric inversion and crossing over does not occur. (2) There is a paracentric inversion with a single crossover with oriented distribution of chromatids yielding all viable gametes. (3) There is a pericentric inversion with no crossover. (b) There is a reciprocal translocation with alternate segregation, giving viable gametes and adjacent segregation, yielding inviable gametes.

26. The F_1 progeny will have a lower fertility than the parents, because the recombinant chromosomes do not have a full genetic complement, and gametes receiving them will be inviable.

27. Examine cells undergoing meiosis with a microscope. An autopolyploid has four sets of identical chromosomes, and in meiotic metaphase these often synapse as tetravalents. An allopolyploid has only two sets of chromosomes, and these can only form bivalents.

28. (a) All gametes from both individuals are euploid. (b) All gametes from the homozygote are euploid, while 50 percent of the gametes from the heterozygote are euploid (Figures 1 and 2). Hence the total frequency of euploid gametes is 75 percent. (c) 50 percent, since one of the parents has only one half of its gametes euploid.

Figure 1

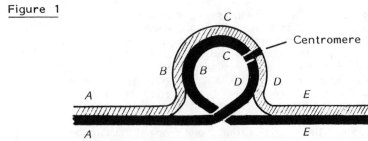

Figure 2

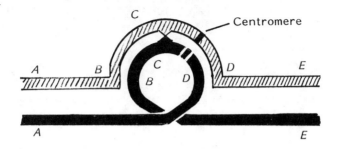

Gene orders in gametes

A B C D E
A D C B E
A D C B A
E D C B E

29. In the tetrasomic, the four chromosomes frequently synapse in twos or as a group of four; in either case the gametes each get two members of each chromosomes. A mating between two disomic gametes, each disomic for the same chromosome, yields a tetrasomic. Thus, self-fertilization of a tetrasomic produces tetrasomic progeny. In the trisomic, two of the chromosomes go to one pole, and one to the other, and the distribution between the poles is random for each triplet. (The mechanism for a two-chromosome—$n = 2$—triploid is illustrated in Figure 6-3 of the text.) Thus, mating will produce some tetrasomics, some normals, and some trisomics, and the strain will not be true-breeding. Also, occasionally the paired chromosomes will be totally lost. Hence it will be the rare zygote that does not suffer from chromosome imbalance.

30. As explained in the previous question, the tetraploid usually produces euploid gametes which, with the haploid gamete of the diploid individual, results in a viable triploid offspring. Gametes from a triploid are usually aneuploid, resulting in inviable offspring.

CHAPTER 7

Genetics of Bacteria and Viruses

CHAPTER SUMMARY

DNA can be transferred between bacteria in three principal ways: transformation, transduction, and conjugation. In transformation free DNA molecules, released from donor cells, are taken up by competent recipient cells; by a recombinational mechanism a single-stranded segment becomes integrated into the recipient chromosome, replacing a homologous segment.

In transduction a generalized transducing phage infects a donor cell, fragments the host DNA, and packages fragments of host DNA into a phage particle. These transducing particles, which contain no phage DNA, can inject donor DNA into a recipient bacterium; by a recombination mechanism the transferred DNA can replace homologous DNA of the recipient, generating a recombinant bacterium called a transductant.

In conjugation, donor and recipient cells pair and a single strand of DNA is transferred by a rolling-circle mechanism from the donor cell to the recipient. During transfer the single strand is converted to double-stranded DNA in the recipient. The transfer process is usually mediated by the transfer genes of the F plasmid. When F is a free plasmid, F becomes established in the recipient as an autonomously replicating plasmid; if F is integrated into the donor chromosome, that is, if the donor is an Hfr cell, only a fragment of DNA is transferred and it can be maintained in the recipient only after a recombination event. About 100 minutes is required to transfer the entire *E. coli* chromosome, but the mating cells normally break apart before transfer is complete. Thus, only a fragment of DNA is normally transferred from an Hfr cell. DNA transfer occurs from a particular point in the Hfr chromosome and proceeds linearly. The times at which donor markers first enter recipient cells--the times of entry--can be arranged in order, yielding a rough physical map of the bacterial genome. This map is circular because of the multiple sites at which an F plasmid integrates into the bacterial chromosome. Occasionally, F is excised from an Hfr cell; aberrant excision, in which one cut is made at the boundary between the F block of genes and the chromosome and the other cut is made in the chromosome, gives rise to F' plasmids, which can transfer bacterial genes. When bacteria are infected with several phage, recombination can occur between phage DNA molecules, generating recombinant progeny. Measurement of recombination frequency yields a genetic map of the phage. A common feature of these maps is clustering of phage genes with related function.

The temperate phages possess mechanisms for recombining phage and bacterial DNA. A bacterium containing integrated phage DNA is called a lysogen, the integrated phage DNA is a prophage, and the overall phenomenon is lysogeny. The integrative exchange occurs between particular sequences called attachment sites. Temperate phages circularize their DNA after infection. Since the phage has a single attachment site, the order of the prophage genes is a permutation of the

order of the genes in the phage particle. The integrated
prophage DNA is stable, but if the bacterial DNA is damaged,
prophage induction occurs, the phage DNA is excised, and phage
development occurs.
   Most bacteria possess transposable elements, which are
capable of moving from one part of the DNA to another. In
bacteria the breakage-rejoining event of transposition does
not utilize sequence homology. Transposition is accompanied by
duplication of a short target sequence at the site of
insertion and duplication of the transposable element itself.
Thus, transposition in bacteria is a replicative process.
Transposable elements frequently carry genes for antibiotic
resistance and can create mutations by integrating within a
host gene and interrupting its continuity. The F plasmid
contains several transposable elements that are also present
in the bacterial chromosome. Recombination between homologous
transposable elements in F and the chromosome are responsible
for formation of Hfr cells.

BOLD TERMS

auxotroph, bacterial attachment site, bacterial
transformation, cohesive end, Col plasmid, colicinogenic
plasmid, colicin, colony, competent, conditional lethal
mutant, conjugation, core, cotransduction, cotransformation,
counterselected marker, cured, drug-resistance plasmid,
excisionase, F plasmid, F' plasmid, heteroduplex analysis, Hfr
cell, high-copy-number plasmid, immunity, inserted, insertion
sequence, integrase, integrated, interrupted-mating technique,
IS element, lysis, lytic cycle, lysogen, lysogenic conversion,
lysogenic cycle, minimal medium, multiplicity of infection,
nonpermissive, nonselective medium, O region, partial diploid,
permissive, phage attachment site, plaque, plaque assay,
plasmid, Poisson distribution, prophage, prophage attachment
site, prophage induction, R plasmid, reciprocal cross,
repressor, selected marker, selective, sex plasmid,
specialized transducing particle, temperate,
temperature-sensitive mutant, terminal redundancy, titer,
transducing particle, transduction, transfer gene,
transformation, transposable element, transposition,
transposon, virulent.

ADDITIONAL INFORMATION

Bacteria and phages have a variety of genetics systems, as
will be shown in Chapter 8, and each uses different modes of
genetic exchange. Rather than review the modes of DNA
transfer, which is presented in a quite straightforward manner
in the text, the number of exchanges required for each mode
will be summarized. First, the reader is reminded that in
meiotic crossing over single crossovers consist of one break
in each chromosome. Recombination in which more than one break
occurs in each chromosome is quite rare.
   In bacterial transformation the recipient DNA must be cut
in two places; that is, transformation requires a double
crossover. In Chapter 8 it will be seen that the exchange is

of a unique sort in that only one strand of the
double-stranded donor DNA enters the recipient cell, and this
single strand exchanges with only a single strand of the DNA
of the recipient. Recombination in Hfr crosses and
transduction also requires a double crossover, because a
double-stranded segment is inserted into the recipient
chromosome. Note also that the exchange is not reciprocal in
either transformation, conjugation, or transduction in that
DNA is always discarded. Transposition also requires two cuts
at the target site.

Prophage integration is a reciprocal process requiring
only a single crossover. Recombination by a single crossover
is a general feature of any system in which two circular DNA
molecules recombine at a single site. Phage-phage
recombination also needs only a single break in both
participating molecules, whether they are linear or circular.
Schematically, though probably not biochemically, phage-phage
recombination resembles chromosome-chromosome recombination in
eukaryotes.

Some additional information can be given about prophage
integration. Whereas the prototype is certainly *E. coli*
phage lambda, both qualitative and quantitative differences
have been observed with other phages. With lambda, insertion
occurs primarily at a single site in the bacterial chromosome,
between the *gal* and *bio* genes. However, some phages
insert their DNA within genes and thereby cause mutations, and
other phages have several major sites of insertion. Many
phages, including lambda can insert in secondary sites, though
the frequency of such insertion is low. With lambda it can be
detected if a bacterial mutant is used in which the lambda
attachment site has been deleted. Integration still occurs at
low frequency (it can be detected by the production of immune
cells, namely, those resistant to infection by lambda) and at
numerous sites scattered throughout the chromosome. Excision
from these sites is often aberrant and frequently produces
transducing particles. This technique of lysogenizing a
deletion mutant has enabled geneticists to obtain a large
number of transducing particles containing bacterial DNA from
many parts of the chromosome. The phage attachment site in
lambda is not within any gene, so that integration does not
cause any phage mutation. This is not always the case. One
interesting phage has its attachment site within its *int*
gene. When integration occurs, the *int* gene is
interrupted; hence, excision of the prophage can never occur!

The integrase protein also mediates exchange between
phage DNA molecules. Lambda contains the attachment site
POP', and integrase is capable of catalyzing an exchange
between two *POP'* sites. Thus, in the absence of all
recombination systems lambda is still able to recombine.
However, only genes on opposite sides of the attachment site
can recombine. Figure 1 illustrates this. Note that a J^-
and a P^- phage can recombine and produce J^+P^+ and J^-P^-
recombinants because genes J and P are on opposite sides of *att*.
However, no recombination can occur between genes F and
J, which are on the same side of *att*. This mode of

Figure 1   Int-promoted recombination

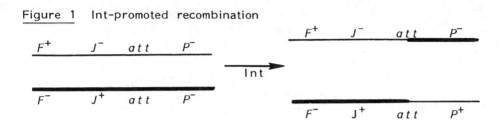

phage-phage recombination, which is known as Int-promoted recombination, has been used to study the mode of prophage integration and excision. All that is needed is a pair of phages carrying different attachment sites and with a mutation in the gene determining normal phage-phage recombination. Experiments are carried out in bacteria defective in bacterial recombination, so that the only recombination that can occur is Int-promoted recombination. *Gal-* and *bio-*
transducing phages do not have the attachment site *POP'*, but instead, because of the aberrant excision, have the attachment sites *BOP'*(*gal*) and *POB'* (*bio*)
thus, a cross between a *gal* and a *bio* phage, namely, *BOP' x POB'*, is precisely the exchange that occurs in prophage excision. The product of the cross is a phage carrying *POP'* and another with *BOB'* (Figure 2). A

Figure 2   Int-promoted recombination between a *gal* and a *bio* transducing particle

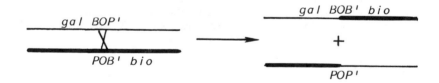

cross between these phages, *POP' x BOB'*, mimics prophage integration. The original *in vitro* experiment in which integrase was first detected used a curious mutant carrying both *att* sites *BOB'* and *POP'*. A circular DNA molecule was used in the assay and Int-promoted recombination was detected by the excision of a DNA fragment between the two *att* sites producing a smaller circle (Figure 3).

Lysogeny has a practical use. In Chapter 10 a bacterial test will be described--the Ames test--for identifying cancer-causing chemicals (carcinogens). Most carcinogens are, for reasons that are not yet understood, able to induce a lambda lysogen to produce phage.

Figure 3 The original *in vitro* assay for integrase

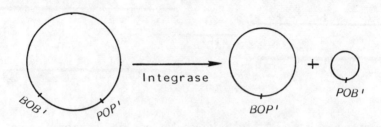

Lysogens are spontaneously induced at a fairly low frequency and in certain bacterial mutants the frequency is lower. Spontaneous induction can be detected quite simply. If one lysogen is mixed with 10^8 bacteria and put on solid medium (exactly as might be done in detecting phage by plaque counting), the lysogen, which is not induced by this process, forms a microcolony (of at most 1000 cells) in the bacterial lawn. Occasionally, late in the development of the colony, a cell will be spontaneously induced, but this does not usually produce a plaque because depletion of nutrients in the growth medium limits cell growth, so the released phage cannot multiply (remember: phage multiply only in growing cells). However, if an inducing agent (in this case, a carcinogen) is in the agar, the original lysogenic cell will be induced to make phage at the time all of the bacteria in the agar begin to grow, so that a plaque will form. If 10^5 lysogens are mixed with the indicator bacteria, the effectiveness of the inducing agent can be measured by the fraction of cells yielding plaques. This test, known as the Devoret test, is a sensitive means of detecting carcinogens, even weak carcinogens, and is being used to examine a very large number of industrial chemicals and food additives.

DRILL QUESTIONS

1. How many DNA molecules are contained in a typical phage particle?

2. What are the three morphological forms of phage particles?

3. How are phage particles usually counted?

4. What is the process by which an infected bacterium releases progeny phage?

5. A particular bacterial mutant cannot utilize lactose as a carbon source. If a phage adsorbs to such a bacterium and the infected cell is put in a growth medium in which lactose is the sole carbon source, can progeny phage be produced?

6. How many plaques can be formed by a single phage particle?

7. A phage adsorbs to a bacterium in a liquid growth medium. Before lysis occurs, the infected cell is added to a large number of bacteria, and a lawn is allowed to form on a solid medium. How many plaques will result?

8. A suspension of a phage is exposed for a long time to a DNase. Will the treatment alter the number of plaques that can be produced by the suspension?

9. If 10^6 phage are mixed with 10^6 bacteria and all phage adsorb, what fraction of the bacteria will not have a phage?

10. In a cross Hfr *leu*$^+$*str-s* x F$^-$*leu*$^-$*str-r*, which markers are selected and counterselected if Leu$^+$Str-r recombinants are desired?

11. If 10^8 phage lambda infect 10^8 *gal*$^+$ bacteria and all phage adsorb, what fraction of the infected cells will yield *gal*-transducing phage?

12. Remember that a lambda prophage is located between the *E. coli gal* and *bio* genes. The *lac* gene is about 10 minutes away from the *gal* gene in the *E. coli* time-of-entry map. Roughly what would be the ratio of *lac*- and *gal*-transducing particles?

13. A phage has gene order *A B C att D E F*. What is the gene order in a prophage?

14. An Hfr strain transfers genes in alphabetical order. Would you expect to obtain F'*y* plasmids lacking gene *x*?

15. An Hfr cell transfers genes in the order *ghi...def*. Which types of F' plasmids could be derived from this strain?

16. An Hfr strain transfers genes in alphabetical order. When using tetracycline sensitivity as a counterselective marker, the number of *h*$^+$*tet-r* colonies is 1000-fold lower than *h*$^+$*str-r* colonies found when using streptomycin sensitivity as a counterselective marker. Explain the difference.

17. An Hfr strain transfers genes in order *a.b.c*. In an Hfr $a^+b^+c^+$*str-s* x F'$a^-b^-c^-$*str-r* mating, will all b^+*str-r* recombinants have received the a^+ allele, and will all b^+*str-r* recombinants also be a^+?

18. Three T4 *rII* deletions *A*, *B*, and *D* have the following properties: *B* x *D* yields *r*$^+$ recombinants but *A* x *B* and *A* x *D* do not. What is the map order of the deletions?

19. Consider the deletions in Problem 18. A mutant yields *r*$^+$ recombinants with *B* and *D* but not with *A*. Locate the mutant with respect to *A*, *B*, and *D*.

ANSWERS TO DRILL QUESTIONS

1. One.

2. Tailed icosahedra, nontailed icosahedra, filaments.

3. In genetic experiments one is concerned with the number of viable phage, so they are counted by plating and counting plaques. However, for physical experiments one may need to know the total number of particles, viable and nonviable. This is not a trivial measurement, though a rough idea (no more than a twofold error) can be obtained by electron microscopy. The best method is to measure the DNA concentration of a solution and divide this value by the molecular weight of the phage DNA molecule, if it is known.

4. Lysis.

5. No, phage can only develop in a metabolizing bacterium.

6. Just one.

7. One. If the bacterium has not lysed, all progeny will be in the same position.

8. No, because DNase cannot penetrate a phage head.

9. Use the Poisson equation. With a phage:bacteria ratio of 1, the fraction uninfected is e^{-1}.

10. Leu$^+$ is selected, and Str-s is counterselected.

11. None, because specialized transducing particles are produced only by lysogens.

12. There will be no *lac*-transducing particles, because the *lac* locus is too far away to be included in a lambda phage head.

13. *D E F att A B C*.

14. No.

15. F'*f*, F'*ef*, ..., but no F'*g*.

16. Genes *h* and *tet* must be very near one another, so selection for the *h* allele of the Hfr tends to select also for the *tet* allele of the Hfr.

17. All receive the allele, but the positions of the exchange determine whether the allele is present in the recombinant.

18. *B A D*.

19. Between *B* and *D* and covered by *A*.

ADDITIONAL PROBLEMS

1. Under what circumstances could a lysate of phage P1 containing transducing particles carry phage lambda?

2. An Hfr transfers genes in alphabetical order. A variant strain V13, known to be lysogenic for a phage XP1, is found that transfers only to gene *e*. That is, genes past *e* never seem to be transferred (at least, *f*-containing recombinants are never formed), and the frequency of transfer of genes *a–d* is less than for the normal Hfr strain. A female strain S132 has the property that transfer from V13 to S132 is normal; that is, time-of entry and plateau values are the same as between a normal Hfr strain and a normal female strain. Suggest an explanation.

3. You have been studying a bacterial species for some years by generalized transduction. Twenty-six genes (*a, b, ..., z*) and thousands of mutations in these genes have been mapped, but because you can only study nearby genes, you have no idea whether the bacterium has one or more chromosomes, and if there is just one, whether it is linear or circular. A single Hfr cell line has been isolated (this fact alone tells you nothing about chromosome structure), and it is found that it transfers all known genes in alphabetical order from gene *a* to gene *z*. (a) What information does this fact alone give you? (b) Reexamination of the transduction data shows that only a few genes can be co-transduced, namely, *a–z*, *c–d*, *h–k*, and *m–n*. What can you now conclude about genome organization in this species?

4. Transposition is a replicative process in which the number of transposable elements increases by one per cell each time transposition occurs. However, cell lines are known in which only a single copy of a transposable element is present. Suggest several possible origins for these cells.

5. Most phages, except for the very small ones, have some genes that are dispensable in all experimental conditions. That is, they can be mutated or even deleted without a significant effect on the phenotype. Given the fact that mutations occur continually, a gene that confers no selective advantage will ultimately be destroyed by repeated mutation. How can you account for the persistence of these "nonessential" genes?

6. Think about Additional Problem 5 when answering these questions. (a) Transducing particles of lambda transduce the *gal* and *bio* genes, two genes that flank the prophage. Would you ever expect to find a lambda transducing particle that can transduce genes further from the prophage than *gal* and *bio*? (b) A specialized transducing particle necessarily lacks phage genes. Do you think that a specialized transducing particle could form a plaque on normal bacteria?

7. When two DNA molecules undergo exchange, theoretically, any number of exchanges could occur. However, what can be said about the number of exchanges if a viable recombinant is to be formed in an Hfr $\times$ F$^-$ mating?

8. Transformation sometimes occurs in nature by spontaneous lysis of a small number of cells and uptake of the released DNA by naturally competent cells. You are examining a number of bacterial species to see which ones can engage in intercellular exchange. A pair of strains of a bacteria that grow in Camembert cheese and have different genetic markers are mixed together. After a period of time the bacteria are plated on selective media and recombinant colonies form. You have reason to believe that transformation is occurring but your colleague thinks that it is a new example of bacterial conjugation. What simple experiment might you do to distinguish these two alternatives?

ANSWERS TO ADDITIONAL PROBLEMS

1. If obtained by infecting a lambda lysogen; the phage DNA would then be no different from any other bacterial DNA.

2. V13 contains an XP1 prophage between e and f. When the prophage is transferred to the female, no repressor is present and phage induction occurs. Thus, recombinants can form only if mating is interrupted before gene f enters the female. Strain S132 probably also contains an XP1 prophage, so the repressor contained in this female strain prevents induction. This phenomenon is called zygotic induction.

3. (a) That fact that all genes are transferred in order suggests quite strongly that the bacterium possesses only a single linkage group and hence that the genome consists of a single chromosome. (b) The transfer order is alphabetical so transfer is presumably alphabetical. Thus, cotransduction of genes a and z shows that the genome is circular, because an "early" and a "terminal" gene must be very near one another.

4. The element may have transposed (1) from a plasmid, which has become lost, (2) from a phage infecting a lysogen for a phage with the same immunity but lacking a transposable element, or (3) from a defective prophage, which been excised without killing the cell. Another possibility is spontaneous excision of the element; this is infrequent but definitely occurs.

5. The key to this question is the statement that the genes are nonessential under experimental conditions. In the laboratory, conditions are usually selected that optimize growth, so certain difficulties encountered by an organism in nature may never be recognized. Clearly, if the phage gene has persisted for millions of years, it must have been essential.

Possibly, it has only recently (months, years?) become nonessential, but a reasonable explanation would be that in nature the gene is essential. Support for this explanation comes from an observation not described in the text. The *E. coli* lambda *b2* gene is certainly nonessential in the laboratory and in fact deletions continually arise, even to the extent of deleting the entire gene. Since lambda is not particularly fastidious about the size of the DNA that it packages and since a smaller DNA molecule takes less time to replicate than a larger DNA, continued cycles of replication of lambda usually results in a population in which most of the phages carry *b2* deletions. Since *b2* deletions are not isolated from natural populations, the *b2* gene must be essential in nature.

6. (a) If a nonessential gene were deleted, the amount of bacterial DNA that could be included in a transducing particle would be greater than that from a normal-sized prophage. Hence, with such a deletion mutant, genes further from the prophage could be transduced. (b) The ability to form a plaques depends on whether essential or nonessential genes are lacking in the transducing particle. With lambda the *gal*-transducing particles always lack the genes for head and tail proteins and hence cannot form plaques. However, the *bio*-transducing particles lack nonessential genes such as *int, xis,* and *red* and are plaque-formers. Some *bio*-transducing particles have such a large *bio* substitution that the essential *N* gene is deleted; these particles do not form plaques.

7. An odd number of exchanges would interrupt the circularity of the chromosome and would undoubtedly prevent replication. Thus, the number of exchanges must be even. Since most female bacteria are not killed by mating, presumably some mechanism exists by which the number of exchanges is always even.

8. Mix the bacteria in the presence of a large amount of a DNase. If recombination resulted from transformation, all transformation would be eliminated, because the DNase would destroy all of the free DNA in the culture. Conjugation would be completely resistant to the enzyme.

SOLUTIONS TO PROBLEMS IN TEXT

1. Genes are injected in the order *his leu trp*. The *met* mutation is the counterselective marker and prevents growth of the male bacteria. The number is small on the His-only medium because crossing over is limited to the regions outside of the *trp* and *leu* genes.

2. Since the donor is $pur^+pro^-his^-$ and pur^+ is selected, we must check for linkage between the pro^- and pur^+ and between his^- and pur^+. With the possible orders *pur pro his* and *pro his pur,* one would expect $pro^- > his^-$, and

his^- > pro^-, respectively. Since his^- (160) > pro^- (25), the first order (pur pro his) is eliminated. The second is possible. With the order pro pur his, nothing can be said about the relative values of pur^- and his^-, because the values depend on the relative spacing. However, with the order pro his pur, one would expect also that since pro is further from pur than his is from pur, then all pur^+pro^- would also be his^-. This is not the case, so the order pro his pur is also inconsistent with the data. The remaining order, pro pur his, must be correct.

3. Think about linkage to answer this problem. Grow P1 on the strain containing the x marker and infect a strain that has a marker A adjacent to x. Select for the A allele present in the strain containing the x marker and test the transductants for the presence of x.

4. Examination of the genes that come in first clearly gives the transfer order b c d. The times of entries are obtained by plotting the number of recombinants of each type versus time and extrapolating back to the time axis. The values are a, 10; b, 15; c, 20; d, 30 minutes. The low plateau value for the d gene is probably the resulting of it being closely linked to the str locus, for in that case, the female d allele (d^-) would be preferred when the female str allele is selected.

5. (a) True, because a enters before c. (b) False, because b enters before c. (c) True, because some crossovers will be between a and b. (d) False; see (c). (e) True, because b is between a and c. (f) True, because the a^- and b^- markers are nearby and on the same DNA molecule. (g) True, because b enters after a.

6. Recombination in an Hfr mating always requires two exchanges, one between the end of the molecule first transferred and the first gene selected for, and the second on the other side of the gene. If a gene is very near the transfer origin, the region in which the particular exchange can occur will be very small and recombinants will be rare.

7. Pro is a terminal marker. F is closely linked to pro and is therefore also at the terminus.

8. 3.5×10^{10}, obtained from the number 352. The numbers fail to follow dilution, because the number 18 is too small to be statistically reliable. The number 2010 is probably too small, because many plaques certainly overlap, causing two plaques to be counted as one.

9. T2 will be turbid, because it fails to lyse the resistant bacteria; T2h will be clear, because it can lyse both the normal and the resistant cells.

10. The prophage gene order is permuted with respect to the phage gene order with the attachment site defining the breakpoint. Since the adjacent genes *f* and *g* have been separated by prophage integration, *att* must be between *f* and *g*.

11. The genes to the left of *att*, in this case, *J*, can be transduced. The nearer they are to *att*, the higher the probability of being included in a *gal*-transducing particle.

12. Genes on opposite side of *att* can be recombined by activity of the integrase system, even when all other recombination systems are absent. This is called Int-promoted recombination. It follows the rules of prophage integration and excision. For example, if one phage carries *BOB'* and the other *POP'*, only Int is required for recombination; however, if the two *att* sites are *BOP'* and *POB'* (that is, a cross between a *gal*-transducing particle and a *bio*-transducing particle), Int and Xis are both needed.

13. (a) The *gal*- and *bio*-transducing particles are excised with the attachment sites *BOP'* and *POB'*, respectively. (b) The cross is *gal BOP'* x *POB' bio*, yielding *BOB'* in a *gal-bio* particle, and *POP'* in a normal phage. (c) No, because the *bio* substitution eliminates the *int* gene. Int-promoted recombination is the only possible recombination mechanism that could carry out the exchange if the other recombination systems were absent.

14. They are only formed by aberrant excision of a prophage. In a lytic infection lambda DNA is not inserted in the *E. coli* chromosome.

15. The time is too short to allow transfer of a terminal gene. Thus, aberrant excision had probably occurred in an Hfr cell, yielding a cell containing an F' plasmid, which carries terminal genes. Thus, the genotype is probably F'z/z⁻str-r, though the colony might have been formed by a reverse mutation of the z⁻ mutation.

16. Somehow the *amp-r* gene has appeared in the lysogens. One possibility is certainly a mutation. However, since the lambda had been grown on an Amp-r host cell, it seems likely that the *amp-r* gene on the host was carried by a transposable element and that transposition occurred to lambda and then to the lysogen.

17. The tetracycline-resistance marker in the phage must have been in a transposable element. Transposition occurred in the lysogen to another location in the *E. coli* chromosome; therefore, when the prophage was lost, the antibiotic resistance remained.

18. The *tet-r* marker in the original F^- strain was carried by a transposable element. In the recombinant this element transposed at a later time to a site near the *lac* gene.

19. (a) Since the ratio of phage to bacteria is 4, by the Poisson distribution virtually all bacteria should have been infected. Thus, the expected number of plaques was 200. (b) The Poisson zero term is 0.5. Therefore, the actual ratio of phage to bacteria was $-\log 0.5 = 0.693$. Thus, $1 - 0.693 = 0.307$ of the phage failed to adsorb.

20. Use the Poisson distribution with $n = 3$ and calculate the fraction of cells infected by two or more phage. This is $1 - P(0) - P(1) = 0.8$. Thus, 80 percent of the 10^8 bacteria (= 8×10^7 bacteria) are infected by two or more phage.

21. (a) These are bacteria to which no phage adsorb. Since the total multiplicity of infection is 5, the fraction of cells which is infected is $e_{-5} = 0.007$. (b) This number includes bacteria to which P4, but not P2, adsorb, as well as the infected cells. The multiplicity of infection of P2 is 3; the fraction of cells not adsorbing any P2 is $e*-3 = 0.05$. Since 0.007 adsorbed neither P2 nor P4, then $0.05 - 0.007 = 0.043$ get one or more P4, but no P2. Hence, $0.043 \times 10^8 = 4.3 \times 10^6$ bacteria get P4 but no P2. (c) These are the bacteria to which P2 adsorb but no P4 adsorb. Using the above reasoning, $e^{-2} = 0.135$ get no P4, and 0.007 get neither. Hence, $0.135 - 0.007 = 0.128$, and 1.28×10^7 adsorb P2 but no P4 (d) These are the bacteria infected by at least one P2 and one P4. The number of cells getting at least one P4 and one P2 is $10^8 - (1.28 \times 10^7) - (4.3 \times 10^5) = 8.22 \times 10^7$.

22. (a) The prophage is transferred by conjugation to a female that lacks the lambda repressor. Hence, the operators are free of repressor and transcription begins. (b) Once this locus is transferred into the female, the female cells will die. Hence, by using the interrupted mating technique, one can determine at what time the numbers of any recombinant (for example, a locus transferred at an early time) decrease.

23. Ultraviolet radiation damages DNA (it interferes with DNA replication and, to some extent, with transcription). T4 obviously does not need any host gene products that must be obtained from the DNA after infection; in contrast, lambda apparently does.

24. A circular DNA molecule lacking genes can be stably established as a plasmid if it has a replication origin that is activated by host enzymes.

25. Plasmid and phage DNA molecules are comparable in size. Phage DNA molecules range in from about 15 million to about

150 million; most however fall in the range, 25-110 million. Plasmids cover the same range, though many are in the 3-20 million range.

26. The segregation of F' plasmids to daughter cells is not perfect and fails in about 1 cell per thousand. If early in the development of a colony a cell divides and one daughter fails to receive F'*lac,* the colony will consist of a mixture of Lac^+ and Lac^- cells. Since cells do not move significantly when multiplying on a solid growth medium, the cells and their progeny will tend to remain on two sides of the colony, so the colony will have two regions of different color on color-indicator media.

CHAPTER 8

Mechanisms of Genetic Exchange

CHAPTER SUMMARY

There are two basic types of exchange--homologous, in which participating base sequences are the same or nearly the same, and nonhomologous, in which the sequences are different. Homologous exchange requires a protein of the type called RecA in *E. coli* for catalyzing strand invasion and pairing of complementary sequences. Nonhomologous exchange utilizes enzymes that are either responsible for joining two particular sequences (site-specific exchange) or for joining a specific sequences with an arbitrarily selected nonhomologous sequence (transposition). Breakage, strand exchange, and rejoining of DNA occurs in all modes of exchange. DNA replication in which gaps are filled and formation of heteroduplex regions are characteristic of homologous exchange.

Formation of single-stranded regions in DNA molecules, a phenomenon called breathing, is important in pairing of homologous regions and is the cause of branch migration. The RecA protein is responsible for the primary association of single- and double-stranded DNA, which occurs without regard for sequence homology; the unwinding activity of this protein allows nonhomologous single- and double-stranded regions to drift back and forth with respect to one another until complementary sequences are matched and lead to pairing. Branch migration extends the size of the paired region. Another enzyme, the RecBC enzyme, is also essential for homologous exchange in *E. coli*, but its function is unknown.

In ascus-forming fungi unusual asci occasionally arise as a result of certain events occurring during meiosis. Crossing over is a reciprocal event, so two molecules with reversed heteroduplex regions form. These aberrant asci, in which genetic markers do not segregate in 4:4 Mendelian fashion, result from the fact that each of the two heteroduplex regions contains different alleles. Since most alleles differ by a single base pair, a mismatched pair of bases is present in each heteroduplex region. These mismatches can be corrected by a repair system. In fungi with ordered asci, if mismatch repair does not occur, asci are formed with two nonidentical pairs of spores. The alleles are still present in a 4:4 ratio, but because of the aberrant order, the asci are called aberrant 4:4 asci or are said to show aberrant 4:4 segregation. When both mismatched pairs are repaired in the same direction (mutant to wildtype or wildtype to mutant), all spore pairs are identical, but the alleles are present in a 6:2 or 2:6 ratio; this phenomenon is called gene conversion or 6:2 segregation. If mismatch repair occurs in only one heteroduplex, an ascus forms in which the members of only one pair of spores have different genotypes; such an ascus is a postmeiotic-segregation ascus and exhibits 5:3 segregation.

A Holliday junction is an intermediate in homologous exchange. Several models suggest means for producing a Holliday junction. The commonly accepted model is the symmetric strand-transfer model, in which pairing occurs by strand invasion and assimilation and an isomerization step creates the crossed strands and a recombinant array of flanking markers. Whereas certain features of crossing over are fairly well understood, a detailed molecular mechanism has not yet been elucidated.

In transposition both the target sequence and the transposable element are duplicated, indicating that replication is part of the process. When a transposable element carried on one DNA molecule transposes to a second DNA molecule, it is likely that a cointegrate, in which two molecules fuse, is an intermediate. Little is known about the molecular mechanism of transposition.

BOLD TERMS

aberrant 4:4, assimilation, asymmetric strand-transfer model, branch migration, breathing, co-conversion, cointegrate, converted, 4:4 segregation, gene conversion, general recombination, heteroduplex region, Holliday junction, homologous recombination, internal resolution site, mismatch, mismatch repair, nonhomologous recombination, postmeiotic segregation, RecA protein, replicon fusion, resolution, site-specific exchange, strand exchange, strand invasion.

ADDITIONAL INFORMATION

In the early 1950s, shortly after the elucidation of the structure of DNA, an experimental approach was taken to understand the molecular mechanism of genetic exchange. It did not take long to recognize that there are many modes of exchange, not only in different organisms but within the same organism. The multiplicity of mechanisms is most evident in *E. coli*, in which one can see transformation, conjugation, transduction, and so forth. However, when one looks closely at *E. coli*, it is found that the multiplicity is even greater than the number of obviously different phenomena. For example, in conjugation it is possible to eliminate about 99 percent of the recombination by a mutation in a critical gene called *recB*. However, some recombination remains, and to eliminate that completely requires mutations in additional genes: *recE*, *recF*, and so forth. It is clear that at least three independent mechanisms are active in the overall processes of conjugation and transduction, all of which depend on an active RecA protein. When phage lambda infects *E. coli*, phage-phage recombination occurs; this exchange does not require the RecA protein, but utilizes two phage gene products called Red and Beta. Phage T4 has a recombination system of its own also. Furthermore, careful analysis of the recombination that occurs with lambda and T4, and during conjugation indicates that these are quite different processes. For some time, it was assumed that although there are many recombination systems, which are more or less important in different organisms, the major recombination system in related microorganisms might be the same and that this system might be the one used in meiotic recombination. Without going into detail it is clear that this is not the case. If there is a relation bwteen meiotic recombination and any system in *E. coli,* it is at best with one of the minor systems. Similarly, the major system in lambda resembles a minor one in *E. coli.* Nonetheless there do seem to be common features in all systems--the enzymes used, the existence of Holliday junctions, and so forth. The general message to be learned is that one can rarely utilize experimental results from one system to understand another system.

Usually students find the subjects of postmeiotic segregation and gene conversion to be somewhat baffling. Actually, they are quite straightforward and easily understood if one takes the time to write out all of the possibilities in a particular ascus. It should be noted though that simple crossing over is not sufficient to generate these "unusual" asci, for their production requires that the position of the heteroduplex region is such that dissimilar alleles are included. We will now present the simplest case, a cross between *A* and *a* in which the heteroduplex includes *A* and *a*. With only one marker recombination is not evident by the presence of new genotypes; however, the formation of unusual asci will make it clear that crossing over has occurred. All possibilities will be written out. Figure 8-12 of the text shows that in the absence of any crossing over the spore order in an ordered ascus is *AAAAaaaa*, that is, 4:4. The result of a crossover that generates *A/a* heteroduplex regions is shown in Figure 1 below. An exchange occurs as shown so that after meiosis I, chromosomes 2 and 3 (arbitrarily numbered) contain the heteroduplex regions. Note that in the absence of any mismatch the spore order is *AA Aa aA aa;* the composition is 4 *A*:4 *a*, with two unmatched spore pairs, so this is an

Figure 1

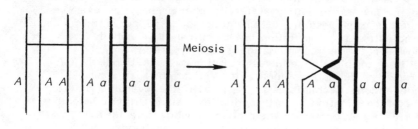

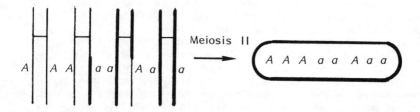

aberrant 4:4 ascus. We now list below the asci that would result with the mismatch repair events indicated.

Correction	Spore order	Ascus type
One *A* to *a* in chromosome 2	AA aa aA aa	3:5
One *A* to *a* in chromosome 3	AA Aa aa aa	3:5
One *a* to *A* in chromosome 2	AA AA aA aa	5:3
One *a* to *A* in chromosome 3	AA Aa AA aa	5:3
a to *A* in both 2 and 3	AA AA AA aa	6:2
A to *a* in both 2 and 3	AA aa aa aa	2:6
a to *A* in 2, and *A* to *a* in 3	AA AA aa aa	4:4 (normal)
A to *a* in 2, and *a* to *A* in 3	AA aa AA aa	4:4 (normal)

The student will find it profitable to go through a similar scheme for the cross *Ab* x *aB*.

DRILL QUESTIONS

1. Are most forms of exchange homologous or nonhomologous?

2. In which exchange process is half of one of the participating DNA molecules discarded?

3. Which exchange processes are unambiguously reciprocal?

4. Does a heteroduplex region that has just formed necessarily contain a mismatched base pair?

5. Compared to a normal Hfr x F^- cross, how many recombinants for a marker transferred early will form if the recipient strain has a *recA*$^-$ mutation?

6. In what way is the replication that occurs in the asymmetric strand-transfer process different from normal DNA replication?

7. If two circular lambda DNA molecules recombine by Int-promoted recombination, what type of molecule results?

8. Consider a double-stranded circle and a linear single-stranded fragment that is homologous to a segment of the circular DNA. The two types of molecules are mixed, heated to a temperature that partially denatures the circle, and then cooled. Examination of the sample by electron microscopy shows some circles with single-stranded appendages. The next day you

prepare fresh samples for electron microscopy and discover that very few circles have appendages; instead, circles with bubbles are seen. What has happened?

9. If a linear lambda DNA molecule engaged in Int-promoted recombination with a bacterial chromosome, the result would be death of the cell. Why?

10. A cross is carried out in *Neurospora* between a wildtype strain (*A*) and a mutant (*a*). Some of the asci are *AAAAaaaa* and others are *AAaaAAaa*. In the formation of which ascus has there been crossing over between the gene and the centromere?

11. How do normal 4:4 and aberrant 4:4 asci differ?

12. A cross *AbD* x *aBd* is carried out with *Neurospora*? Among several types of asci the following are observed: (1) *AbD, AbD, Abd, Abd, aBD, aBD, aBd, aBd;* (2) *AbD, AbD, ABd, ABd, aBD, abD, aBd, aBd;* (3) *AbD, AbD, ABd, ABd, aBD, aBD, aBd, aBd.* Classify these asci, and identify the pairs in which conversion has occurred.

13. How many mismatch corrections occur in a 6:2 asci, and what can be said about the direction of the correction?

14. What enzyme is responsible for strand assimilation, and could an enzyme like DNA polymerase III suffice?

15. All types of exchange processes (except for Int-promoted exchange) are terminated by the same enzymatic event. What is that event?

16. With reference to question 8, what happens that is different in Int-promoted exchange?

17. Which types of recombination necessarily require two exchanges?

18. Name two proteins capable of bringing two DNA molecules together.

19. Homologous recombination requires a RecA-type protein, but RecA brings together molecules independently of their homology. Explain where homology enters the process.

20. In early chapters we learned that recombination frequency can never exceed 50 percent. How is this explained by either the Holliday or asymmetric strand-transfer models?

21. Name five genetic phenomena mediated by transposable elements.

ANSWERS TO DRILL QUESTIONS

1. Homologous.

2. Bacterial transformation.

3. Site-specific, probably phage-phage exchange, and probably meiotic recombination.

4. Not if it forms between strictly homologous regions. In drawing models of exchange one usually draws the exchanges between genetic markers, so that the heteroduplex region contains the two markers. However, exchange occurs at random and more often than not it will occur in regions of complete homology. Such exchanges will of course be undetectable and are not particular interesting from a genetic point of view, but they nonetheless occur. In carrying out physical experiments to study exchange, these nonrecombinational exchanges must of course be considered.

5. A *recA*⁻ mutation eliminates recombination.

6. No initiation protein or replication origin is needed.

7. A circular dimer.

8. The single-stranded appendage has branch-migrated and been assimilated.

9. If a linear molecule (lambda) recombines with a circular molecule (*E. coli*), the circle becomes linearized--that is, broken.

10. The one with the gene order *AAaaAAaa*.

11. In a normal ascus both members of each spore pair are identical; in an aberrant 4:4 ascus *two* spore pairs have different members, though the overall ratio of alleles for each gene is 4:4.

12. (1) is a normal ascus; no conversion events. (2) is a 5:3 ascus; in the second pair of spores a *b*-to-*B* conversion has occurred; (3) is a 6:2 ascus; in the second and third pairs of spores a *b*-to-*B* conversion has occurred.

13. There are two corrections and both are in the same direction, that is, both *x* to *X*, or both *X* to *x*.

14. A polI-type enzyme. No, because the enzyme must be able to act at a nick and must have displacement activity, neither of which is possessed by polIII.

15. Ligation by a DNA ligase.

16. Integrase carries out the ligation itself.

17. Transduction and bacterial conjugation.

18. RecA and integrase.

19. The initial interaction between two DNA molecules does not require homology. A stable interaction requires that the two molecules drift until complementary sequences are paired.

20. In the resolution step there is a probability of 50 percent of cuts that would produce recombination and 50 percent of cuts that yield the nonrecombinant configuration of markers.

21. Replicon fusion, deletion, inversion, transposition, and mutation.

ADDITIONAL PROBLEMS

1. A Lac⁻ *Pneumococcus* is treated with DNA from a Lac⁺ strain and plated on a nonselective color-indicator medium in which Lac⁺ and Lac⁻ cells produce red and white colonies, respectively. What types of colonies will form on the medium?

2. *In vitro* studies of Int-promoted recombination have shown that the DNA molecule containing *POP'* must be supercoiled. The other participant need not be supercoiled. Whereas it has not been proved, the explanation given for the requirement for supercoiling is that it eliminates the possibility of a potentially lethal exchange. What is the nature of this exchange?

3. Transposition is a type of nonhomologous exchange, suggesting that matching of sequences is not part of the process. However, there seems to be some sequence effect in that transposition occurs to some sites more than to others. The mechanism of this effect is unknown. Suggest what type(s) of things might account for this effect.

4. The exchange mechanism needed in transduction and in Hfr × F⁻ crosses was not discussed in the text. Would you expect them to be like transformation or like each other?

5. In a Hfr *leu⁺str-s* × *F⁻leu⁻str-r* mating in which the *leu* gene is transferred early, there are roughly 40 *leu⁺str-r* recombinants per 100 Hfr cells. If the F⁻ cells are also *recA⁻*, the number of such recombinants is about 10^{-6}. This reduction is true for all early markers. However, if one studies markers transferred very late, for which the recombination frequency is generally much lower than 40 per 100, it is observed that a *recA⁻* mutation in the recipient cell does not alter the recombination frequency at all. Explain this lack of reduction, which is a general feature of the transfer of late markers to a *recA⁻* strain.

6. Consider a double-stranded circular DNA molecule and a single-stranded fragment, half of which is homologous to a sequence in the circle and half of which is totally different. The molecules are mixed, heated to a temperature at which the circle is partially denatured, and then cooled slowly. The sample is examined by electron microscopy and circles with single-stranded appendages of various length are seen. The next day fresh samples are prepared for microscopy. No circles with appendages are seen. Instead, all circles appear to be normal double-stranded circles. Free single strands are seen also. Explain these observations. (Compare to Drill Question 8).

7. A lambda phage can be crossed with a *gal*-transducing particle even though there is a region of homology. For instance, there is no difficulty in detecting P^+Q^+ recombinants in a cross between a P^-gal particle and a Q^- normal phage. However, if two markers are very near but definitely not within the region of homology, the recombination frequency between them is depressed compared to that with normal phages. How would either Holliday or Meselson and Radding explain this difference?

8. Suppose that phage mutants are isolated that show decreased ability to recombine. All mutants fall into two complementation groups, *recX* and *recY*. The known genes of the phage have the map order d e f g h i j k l m n. Crosses are performed with Rec⁻ mutants. The data are shown in Table 1. (a) What properties of the *recX* and *recY* systems are indicated by these results? (b) What is the probable relation between the physical spacing of adjacent genes?

Table 1

Genotypes of parental phages in crosses	Genotype of recombinant	Recombination frequency, percent	
		recX phages	*recY* phages
d^-, e^-	d^+e^+	0.0001	1
f^-, j^-	f^+j^+	0.0001	4
j^-, m^-	j^+m^+	1	2
j^-, n^-	j^+n^+	1	3
k^-, m^-	k^+m^+	1	1
k^-, l^-	k^+l^+	1	0.01
j^-, k^-	j^+k^+	0.0001	1
l^-, m^-	l^+m^+	0.0001	1

ANSWERS TO ADDITIONAL PROBLEMS

1. White (nontransformant, or a cell in which the lac^+ DNA was integrated and in which there was a conversion from + to -), red (one lac^+ strand integrated and a conversion from - to +), and sectored (one lac^+ strand integrated and no conversion).

2. If a linear DNA molecule were to exchange with a bacterial chromosome, the circular chromosome would be broken (See Drill Question 4). A requirement that the incoming phage, which contains POP', be supercoiled and hence circular eliminates the possibility of such a disastrous event.

3. An effect of sequence on the probability of the first breakage effect seems most likely.

4. They would not be like transformation since a double-stranded DNA must enter the cell and in transformation only a single strand enters. They are probably similar to one another since double-stranded DNA is involved, two exchanges are needed, and RecA protein is required.

5. The Hfr cell is $recA^+$, so the $recA$ gene is also transferred at some time to the recipient. Once this happens, recombination can occur.

6. Branch migration causes the base-paired region to drift back and forth. However, migration cannot occur across the nonhomologous region, so the single strand can never be completely assimilated. As the branch migrates back and forth, occasionally migration will extend to the homologous end of the single strand, at which point the single strand will dissociate completely from the circle. The single strand cannot penetrate the circle again at a low temperature, so ultimately all circles lose the partially assimilated strand.

7. Both would say the same thing. Branch migration cannot occur across the nonhomologous region, which limits the size of the region in which the initial association between the molecules can occur. Reducing the size of this region will reduce the probability of interaction and hence the recombination frequency.

8. (a) $RecX$ is a gene for general recombination; $recY$ catalyzes site-specific exchange between genes k and l. (b) All genes are roughly equally spaced except k and l, which are very near.

SOLUTIONS TO PROBLEMS IN TEXT

1. Homology and a RecA-type protein are required for

CHAPTER 8

transduction, transformation and bacterial conjugation. Meiotic recombination requires homology and presumably a RecA-type protein also. Transposition and site-specific exchange require neither. One way that transposition is usually detected is by observing the genetic change in a mutant having a nonfunctional RecA protein.

2. Prophage integration and excision, formation of an F', formation of an Hfr, and meiotic recombination are reciprocal events, because neither genetic information nor DNA is lost. Meiotic recombination may appear nonreciprocal when conversion occurs, but the physical exchange remains reciprocal. Transformation is nonreciprocal, because one DNA strand is discarded and because only a portion of the strand that enters the bacterium is ever integrated. Transduction is nonreciprocal, because material from both ends of the double-stranded DNA fragment are discarded.

3. The target sequence and in prokaryotes the transposable element itself. In eukaryotes some transposable elements are actually excised and move, without leaving a copy at the original site.

4. 3 and 5. Molecule 1 is impossible because the two strands shown as forming a short double-stranded segment at the upper left have the same polarity and could not base pair. Molecule 2 is not possible because the short double-stranded segment is the upper part of the right fork consists of two segments having the same polarity. Molecule 4 has the same deficiency as molecule 2. In molecule 6 the short and long single-stranded appendages in the right fork would form a short double-stranded appendage, because they are adjacent and in register.

5. Postmeiotic segregation, for in these asci pairs of spores carry different alleles and each member of the pair is derived from one strand of the double-stranded meiotic product. Thus, each strand carries a copy of a different allele, which is one definition of a heteroduplex region.

6. Branch migration, for the initial point of joining is not usually the point at which resolution of the Holliday junction occurs.

7. Mismatch repair and by copying one parental strand during the assimilation process.

8. It Increases recombination frequency by providing breaks, gaps, and free ends, which facilitates pairing. Also, synthesis of the RecA protein is stimulated, but that was not described in this book.

9. I are nonrecombinant. II are recombinant without conversion. III are recombinant with an E-to-e conversion.

10. The exist of aberrant 4:4 asci shows that heteroduplex regions are produced in the exchange process. The remaining types all arise when mismatch repair of a heteroduplex regions occurs, since they are absent, the mismatch repair system must be absent. This situation can be observed with mutants that eliminate activity of the mismatch repair system.

11. Immediately after invasion, the recipient has a mismatched base pair. For low-efficiency markers mismatches are usually corrected to the recipient genotype. For high-efficiency markers either mismatch repair fails to occur or correction is primarily to the donor genotype.

12. Two circles joined by crossed strands, that is, a figure 8. The mechanism of formation of the figure-8 molecule is shown in Figure 2.

Figure 2 Formation of a figure-8 molecule from two circles

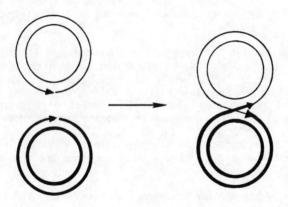

CHAPTER 9

Gene Expression

CHAPTER SUMMARY

The flow of information from a gene to its product is DNA to RNA to protein. The properties of the different protein products of genes are determined by the sequence of amino acids of the polypeptide chain and by the way the chain is folded. Each gene is usually responsible for the synthesis of a single polypeptide. Exceptions are found among the small viruses in which three different mechanisms--multiple reading frames, differential RNA splicing, and polyprotein cleavage--enable several proteins to be made by a single tract of DNA bases.

Gene expression begins by the enzymatic synthesis of an RNA molecule that is a copy of one strand of the DNA segment corresponding to the gene. This process is called transcription, and the enzyme responsible for it is RNA polymerase. The substrates of RNA polymerase are ribonucleoside triphosphates. The chemical reaction catalyzed by the enzyme is the same as that used in DNA synthesis. A difference between DNA and RNA polymerases is that RNA polymerase does not need a primer to initiate synthesis. Transcription is initiated when RNA polymerase binds to a promoter sequence. Each promoter consists of several subregions, of which two are the polymerase binding site and the polymerization start site. Polymerization continues until a termination site is reached. The product of transcription is an RNA molecule called a primary transcript. In prokaryotes this molecule is used directly as messenger RNA (mRNA) in polypeptide synthesis. In eukaryotes the primary transcript is processed: noncoding sequences called introns are removed and the termini are modified by formation of a 5' cap and addition of a 3'-poly(A) sequence. Some primary transcripts do not contain mRNA sequences but are cleaved to produce tRNA and rRNA molecules.

After mRNA is formed, polypeptide chains are synthesized by translation of the mRNA molecule. Translation is the successive reading of the base sequence of an mRNA molecule in groups of three bases called codons. There are 64 codons; 61 correspond to the 20 amino acids, of which one (AUG) is a start codon. The remaining three codons (UAA, UAG, UGA) are stop codons. The code is highly redundant, in that many amino acids correspond to several codons. The codons in mRNA are recognized by tRNA molecules, which have a three-base sequence complementary to a codon and called an anticodon. When used in polypeptide synthesis, each tRNA molecule possesses a terminally bound amino acid. The correct amino acid is attached to each tRNA species by specific enzymes called aminoacyl tRNA synthetases. Polypeptide synthesis occurs on ribosomes, particles containing several RNA molecules and numerous proteins. Synthesis begins by binding of a ribosome and an mRNA molecule and by occupation of two sites on the ribosome with the charged initiating molecule tRNA^{fMet} and a charged tRNA molecule carrying the second amino acid of the polypeptide chain. The amino acids are joined, the tRNA^{fMet} is removed, and the mRNA molecule and ribosome are displaced with respect to one another to allow another charged tRNA molecule to bind. This process continues until a stop codon is reached; at this point the polypeptide is released. Several ribosomes can translate an mRNA molecule simultaneously, forming a polysome. In prokaryotes, translation often begins before synthesis of mRNA is completed; in eukaryotes this does not occur because mRNA is made in the nucleus, whereas the ribosomes are located in the cytoplasm. Prokaryotic mRNA molecules are often polycistronic, encoding several different polypeptides. Translation proceeds sequentially along the mRNA molecule from the start codon nearest the ribosome binding site, terminating at stop codons and reinitiating at the next start codon. This is not possible with eukaryotes, because only the AUG site nearest the 5' terminus of the mRNA can be used to initiate polypeptide synthesis. Eukaryotic mRNA is monocistronic.

BOLD TERMS

A site, acylated, amino terminus, amino acid attachment site, aminoacyl site, aminoacyl tRNA synthetase, anticodon, backbone, cap, carboxyl terminus, charged tRNA, colinearity, chain elongation, chain initiation, chain termination, coding sequence, coding strand, codon, consensus sequence, coordinate regulation, core enzyme, coupled transcription-translation, disulfide bond, domain, EF-G, elongation, exon, 40S ribosome, frameshift mutant, gene expression, gene product, genetic code, initiation, initiation factor, intergenic complementation, intervening sequence, intragenic complementation, intron, messenger RNA, metabolic pathway, mischarged tRNA, mRNA, negative complementation, one gene-one polypeptide hypothesis, overlapping genes, P site, palindrome, peptide bond, peptidyl site, peptidyl transferase, polar mutation, poly(A) tail, poly(A)-addition site, polycistronic mRNA, polymerization start site, polypeptide chain, polyprotein, polyribosome, polysome, Pribnow box, primary transcript, processing, promoter, promoter mutation, promoter recognition, R group, reading frame, recognition region, redundant, release factor, ribosome binding site, RNA polymerase, RNA processing, RNA splicing, rRNA, sense strand, Shine-Dalgarno sequence, 60S ribosome, 70S initiation complex, 70S ribosome, sigma subunit, small ribonucleoprotein particle, snurp, spacer sequence, start codon, stop codon, subunit, TATA box, template, termination, 30S and 50S ribosome, 30S preinitiation complex, transcription, translation, triplet code, tRNA, uncharged tRNA, wobble hypothesis.

ADDITIONAL INFORMATION

By the time the student has completed Chapter 9 there is no doubt that he or she will be overwhelmed by the complexity of the scheme for production of a finished functional protein from a gene; the student, who does additional and more advanced reading, will also recognize the elegance of the process and the various ways that the frequency of errors is kept down. It is worth mentioning, though it is not an encouraging thought, that the process is even more complex than has been presented in the chapter, because many details and additional protein factors have not been mentioned (this will not be done here). However, the basic mechanism is a simple one: a base sequence in a DNA molecule is converted to a complementary base sequence in an intermediate--mRNA--and then the base sequence in the mRNA is converted to an amino acid sequence of a polypeptide chain. Both of these steps, which admittedly have a multitude of substeps, utilize the simplest of principles: (1) the rules of base pairing provide the base sequence of the mRNA, and (2) a two-ended molecule (tRNA), which can bind an amino acid at one end and can base pair with RNA bases at the other, converts each set of three bases to one amino acid. Various recognition regions are needed to be sure that the right base sequence is read and that the right amino acid is put in the appropriate position in the protein. How are these things accomplished? As is always the case when utilizing the information in nucleic acid molecules, base sequences provide the information for the first process. That is, specific sequences in the DNA are recognized as the beginning (promoter) and end (transcription termination site) of a gene, and these sequences are recognized by an enzyme (RNA polymerase) that makes the copy of the gene that is used by the protein-synthesizing machinery. To ensure that the correct amino acid sequence is assembled, a specific base sequence (AUG) in the mRNA molecule is used to tell the system where to start reading, another sequence (a stop codon) defines the end of the polypeptide chain, and particular recognition sites in tRNA enable specific enzymes (the aminoacyl tRNA synthetases) to connect an amino acid to the sequence in the tRNA (the anticodon) that is complementary to the sequence in the mRNA (a codon) specifying the amino acid.

An essential feature of the entire process of gene expression is that both DNA and RNA are scanned by molecules that move in a single direction. That is, RNA polymerase moves along the DNA as it polymerizes nucleotides, and the ribosome and the mRNA move with respect to one another as different amino acids are brought in for covalent linking. Such movement is quite unusual in enzymatic systems, or in chemical reactions in general. In a typical chemical reaction an enzyme attaches to its substrate, reaction or molecular rearrangement occurs, and the enzyme releases the product, ready to find another substrate molecule. However, in this situation RNA polymerase does not let go of the DNA but moves to the next base in line as soon as polymerization has occurred. The mechanism for this movement is not known, but there are some good guesses. Probably the best one would be that RNA polymerase, which is a huge multisubunit protein probably changes shape after it has bound to a particular base. Such shape changes are common with enzymes and in multisubunit proteins can be quite large. It is quite possible that the change is such that during the act of actually making the phosphodiester bond, the region of the enzyme that recognizes the base is put very near the next base. When polymerization is complete, the enzyme regains its shaped using in nucleotide binding and then binds to the nearest base, which is the next in line. Such a mechanism is pure speculation, but the student should know that this is the kind of thing that happens frequently in reactions of this type (so-called processive reaction). Movement also occurs in protein synthesis, namely, as the ribosome moves to successive codons. In this case, the movement is the result of changing tRNA-ribosome interactions. That is, after two amino acids are linked, the last-linked amino acid is removed from the tRNA and the tRNA undergoes a small change in shape that makes it unable to bind to the ribosome. How the other tRNA jumps from one binding site to the next is not known, but a shape change is presumably involved. One can imagine a shape change in the ribosome that would bring the tRNA in the P site to the A site. Again this is highly speculative and the so-called translocation mechanism is poorly understood; however, something of this sort certainly happens.

Ribosomes are themselves remarkable complicated, having a very large number of components. Some of the components are simply structural but others certainly have enzymatic activity. The peptide bond is formed by the peptidyl transferase complex, whose active site is formed from regions of several ribosomal proteins. Other proteins form the tRNA binding sites, and a third class is involved in an unknown way with termination and release of finished polypeptide chains and with translational fidelity. The functioning of the ribosome is still somewhat of a mystery.

A question that must have entered the minds of the reader is: what is so special about the third codon position that it is less important in base pairing? The real question should probably be: how can three bases in a codon manage to base-pair stably with three bases in an anticodon? This number is quite small. For example, a double-stranded hexanucleotide is exceedingly unstable at room temperature, and two DNA molecules having only one complementary trinucleotide sequence will not form a pair that can survive more than a few microseconds. Again one must resort to the shape of the ribosomes to explain this stability. Clearly, the codon-anticodon interaction is stable when the mRNA is bound to a ribosome. It is not known why this is so, but a general statement can be made: both the tRNA-ribosome and codon-anticodon interactions are by themselves weak, but together they are strong. This is akin to the explanation for the joint roles of hydrogen bonding and hydrophobic interactions in DNA stability. Recall that if either hydrogen bonds or the hydrophobic interaction is broken down, the strands of DNA will come apart. Cooperative interactions of this sort are common in biochemical systems. With this in mind, the lack of participation of the third codon position becomes less mysterious. When tRNA and mRNA are both bound to a ribosome, this base and the corresponding base in the anticodon must be situated such that their interaction contributes very little to the stability of the tRNA-mRNA complex. Some pairing is necessary to complete the stabilization (because any base in the anticodon will not suffice), but a weak interaction from a nonstandard base pair will suffice. Thus, one can argue that the shape of the ribosome is the true determinant of the redundancy of the code.

The final point worth making is about the differences between the overall process of protein production in eukaryotes and prokaryotes. Perhaps one should first consider the similarities, which are so great that one would never even conceive of the possibility that the systems evolved independently. Without examining details of the differences, the most striking feature of the eukaryotic system is that both the informational component (the RNA) and the final product (the protein) are modified. That is, introns are excised from the eukaryotic primary transcript and the polypeptide chain is frequently cleaved, rejoined, and chemically modified. Why processing occurs in almost every case in eukaryotes and only occasionally in prokaryotes (for example, in the production of rRNA and tRNA) is not known. However, a suggestion for the origin of some introns has been given in Section 9.3. When one thinks about evolution, it is interesting to note that the complexities of eukaryotic transcription and translation are found both in the free-living unicellular and the higher eukaryotes, suggesting that they evolved in a primitive organism. Whether one considers that prokaryotes preceded eukaryotes or vice versa, it is interesting that such changes should be associated either with the evolution of or loss of the nuclear membrane.

DRILL QUESTIONS

1. What can be said about the relative amounts of each base in RNA?

2. What uracil-containing molecule is a precursor for RNA synthesis?

3. What are the three stop codons?

4. What is the principal start codon and to what amino acid does it correspond?

5. How do prokaryotes and eukaryotes differ in the methods for selecting an AUG site as a start codon?

6. Which two amino acids can be placed in a polypeptide chain at an AUG codon in prokaryotes. How does the situation differ in eukaryotes?

7. What two amino acids have only one codon?

8. What change does RNA polymerase undergo shortly after initiation of RNA synthesis?

9. Poly(U) encodes polyphenylalanine. If a G is added to the 3' terminus, the polyphenylalanine will have another amino acid at its terminus. What is the amino acid, and will the

same amino acid be added at the terminus of polyphenylalanine if the G is added to the 5' terminus?

10. Both rRNA and tRNA precursors are spliced in forming the finished RNA molecules. However, only in forming mRNA is the excised RNA called an intron. What feature defines an intron?

11. If a DNA-RNA hybrid were made, would the strands be parallel or antiparallel?

12. In protein synthesis is the final amino acid added to the amino or the carboxyl end of the polypeptide?

13. Which of the following is the normal cause of chain termination: (1) The tRNA corresponding to a chain-termination triplet cannot bind an amino acid; (2) there is no tRNA with an anticodon corresponding to a chain termination triplet; (3) mRNA synthesis stops at a chain termination codon?

14. How many nucleoprotein subunits are there in a prokaryotic and a eukaryotic ribosome?

15. How many RNA molecules are in a prokaryotic and a eukaryotic ribosome?

16. How many tRNA molecules respond to AUG?

17. What amino acids are bound to charged $tRNA^{Leu}$ and to seryl-$tRNA^{Leu}$?

18. How many amino acids are linked together at the time the ribosome and mRNA first move with respect to one another?

19. What is meant by heterogeneous nuclear RNA?

20. At which ends of eukaryotic mRNA are the cap and the poly(A) tail?

21. Reconcile the facts that in eukaryotes a primary transcript can have the coding sequences for two proteins yet only one protein can be translated from a eukaryotic mRNA molecule.

22. In one organism eight different mutants are isolated that induce a requirement for histidine biosynthesis. What does this result tell you about histidine biosynthesis in this organism? (Think carefully about this one.)

ANSWERS TO DRILL QUESTIONS

1. Nothing. They can have any value.

2. Uridine triphosphate (UTP).

3. UAA, UAG, UGA.

4. AUG; methionine.

5. In prokaryotes the start codon is the first AUG after a Shine-Dalgarno sequence; in eukaryotes the start codon is the nearest AUG to the 5' end of the RNA.

6. Methionine and formylmethionine; in eukaryotes, only methionine.

7. Methionine and tryptophan.

8. Loss of the sigma subunit.

9. With G at the 3' terminus the poly(U) will terminate with UUG and the polypeptide will end with leucine; if it is at the 5' terminus, the poly(U) will start with GUU, so the polypeptide will initiate with glycine.

10. An intron interrupts a coding sequence.

11. Antiparallel.

12. Carboxyl end.

13. (2).

14. Two, in each case.

15. Prokaryotic, three; eukaryotic, four.

16. In both prokaryotes and eukaryotes, two: the tRNA^{Met} and the initiating tRNA.

17. Leucine to tRNA^{Leu} and seryl to seryl tRNA^{Leu}.

18. Two, those in the A and P sites.

19. The collection of primary transcripts and partially processed RNA in the nucleus.

20. The cap is at the 5' end; the poly(A) tail is at the 3' end.

21. Only the coding sequence initiating from the AUG nearest the 5' terminus of the RNA can be translated.

22. Nothing whatsoever, since you do not know the number of complementation groups.

ADDITIONAL PROBLEMS

1. The rate of initiation of new RNA strands *in vitro* is greater with a superhelical template than with a linear DNA molecule or with a nicked circle. Propose an explanation for this phenomenon.

CHAPTER 9   131

2. *In vitro* protein synthesis is being carried out using (AUG)$U_{90}G_{90}$ as an mRNA molecule. The system has been obtained from bacteria known to carry a temperature-sensitive mutation that blocks protein synthesis *in vivo* within a few seconds after the temperature is raised. *In vitro,* if the temperature is raised a few seconds after initiation, the longest polypeptide found is Met-Phe$_{30}$, though some with fewer phenylalanines are occasionally observed. Suggest a defect in this mutant system.

3. Suppose the mRNA in Problem 2 had not started with AUG, and the explanation for the defect had been the same. How would the experimental result have been different?

4. What anticodons probably corresponds to the codon UGG?

5. A biochemical pathway is being analyzed by isolating mutants. All mutants required compound H to grow. With each of the mutants four compounds--A, B, C, and D--were tested for their ability to support growth of the mutant. The results were the following: Mutant 1 would grow if D was supplied, but not with the others. Growth of mutant 2 was supported by either B or D. Mutant 3 would grow only if H were added to the medium. Mutant 4 would grow if either A, B, or D were provided. Mutant 5 would grow with either B or D, or if A and C were provided together (mutant 5 is obviously a double mutant). What is the pathway, and at which step is each mutant blocked? (Hint: the pathway starts with a reaction between two substances and one mutant is a double mutant.)

6. A series of mutants in *Neurospora* requires phenylalanine (P), tyrosine (T), or both for growth. Each mutant accumulates a particular intermediate in the biochemical pathway for synthesis of these amino acids. The table below shows the ability (+) or inability (-) of the individual mutants to grow when the compounds indicated (designated by letters) are added to the medium; the precursors to phenylalanine and tyrosine that accumulate in each mutant are also given. Derive the biosynthetic pathway for the two amino acids. Note: the pathway has one branch, and mutants that block certain steps in the pathway have not been found.

Mutant	A	B	C	T	P	T + P	Accumulates
1	+	+	+	-	-	+	E
2	+	+	-	-	-	+	F
3	-	+	-	-	-	+	D
4	-	-	-	+	-	+	B
5	-	-	-	-	+	+	B

7. RNA polymerase lacks the nuclease activities of DNA polymerase. Name two activities possessed by RNA polymerase and not by DNA polymerases.

8. Without carrying out any sequencing analyses what indication might you have that all promoters do not have the same overall base sequence?

9. Consider both the code as we know it and the wobble hypothesis, what is the minimum number of tRNA moleules needed to recognize the 61 codons corresponding to amino acids?

10. What three requirements must be met if an aminoacyl synthetase is to charge a tRNA molecule?

11. Consider a segment of DNA in the very early part of a long protein. This small segment can tolerate many amino acid changes without altering the activity of the protein. The sequence of nucleotides is the following: TCGTAGAGGG|GCA-ATGAGC AATACCCCGA.... A mutation occurs in which two G's get inserted at the |. This change inactivates the protein, because the reading frame is altered by the two bases. However, if another mutation occurs that changes the underlined G to a C, an active protein (though not necessarily the same one) is made. Explain.

ANSWERS TO ADDITIONAL PROBLEMS

1. RNA polymerase must come into contact with bases, which means that some unwinding of the DNA is necessary. Recall from Chapter 5 that supercoiled DNA is partially underwound and that single-stranded regions have a longer lifetime than in nonsupercoiled DNA. Thus, at any given point in the DNA the bases in supercoiled DNA will be accessible to the polymerase for a greater fraction of the time.

2. There seems to be nothing wrong with the protein synthetic system itself. However, the system is unable to translate from the G portion of the mRNA. The simplest explanation is that no charged tRNA is present that can read the GGG codon. This could result from a temperature-sensitive tRNA molecule or aminoacyl synthetase.

3. The system would have to be one that did not need an initial AUG. In that case, translation would not always start at the end of the mRNA. This would have two consequences. First, the polypeptide would not start with Met, but this is obvious. Second and more important, some of the polypeptides would contain fewer than 30 phenylalanines, when translation began within the mRNA. Third, the reading frame would sometimes be different, so the final codon would sometimes be either UUG or UGG. Therefore, some of the polypeptides would have terminated with leucine or tryptophan.

4. The base that could pair with the G in the third position could be C or U. If it were U is would also pair with the codon UGA, which is a stop codon. Thus, it is not U and must be C. Therefore, the anticodon is 3'CCA.

5. See Figure 1.

Figure 1

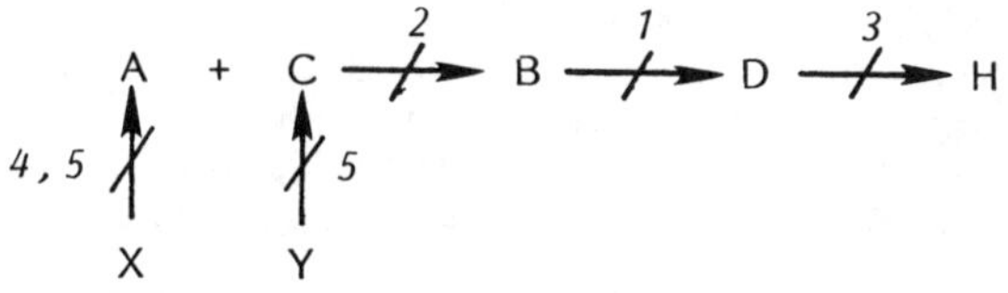

$$A + C \xrightarrow{2} B \xrightarrow{1} D \xrightarrow{3} H$$

with $4,5 \updownarrow$ on A to X, and $\updownarrow 5$ on C to Y.

6. The reaction sequence is E-C-F-A-D-B with a branch at B, one branch leading to tyrosine and the other to phenylalanine. The reactions blocked by the mutations are : 1, E-C; 2, F-A; 3, D-B; 4, B-tyrosine; 5, B-phenylalanine.

7. It is self-priming (that is, it does not need a primer), and it can unwind DNA without the need of a helicase.

8. The rates of initiation of synthesis of all mRNA molecules in a particular organism differ from one gene to the next.

9. Thirty-two.

10. The amino acid must be present in an activated form (actually an amino acid-AMP complex), the synthetase must be able to bind to the tRNA, and the three-dimensional structure of the tRNA molecule must be such that the CCA (3') terminus is positioned in the active site of the synthetase.

11. The G-to-C mutation creates a UAG stop codon. Thus, the AUG just downstream and in the correct reading frame can be used to start in-frame synthesis again.

SOLUTIONS TO PROBLEMS IN TEXT

1. Ribonucleosides triphosphates, RNA polymerase, and one strand of DNA, respectively.

2. Each has a 3'-OH group. A primary transcript has a 5'-triphosphate; prokaryotic mRNA has a triphosphate, and eukaryotic mRNA usually has a cap.

3. In an overlapping code, a base change would affect three codons, so that often more than one amino acid would be changed. (One is possible because of the redundancy of the code.) Thus, an overlapping code was eliminated from consideration.

4. The start codon is complementary to the DNA base sequence TAC (bases 4, 5, 6). Note that an assumption must be made about the direction of reading the strand. The convention is to write the coding strand so that the mRNA can be read from left to right; that is, the coding strand is written with the

3' terminus at the left. The amino acid sequence is
NH$_2$-Met-Pro-Leu-Ile-Ser-Ala-Ser.

5. If the average protein has a molecular weight of 50,000, it has about 454 amino acids, which corresponds to 3(454) = 1362 base pairs, or a DNA molecular weight of 8.25 million. Dividing the molecular weight of the bacterial DNA by this value, yields about 3000 genes, or 3000 polypeptides.

6. Codons are read from the 5' end of the mRNA, and the first amino acid in a polypeptide chain is at the amino terminus. If the G were at the 5' end, the first codon would be GAA (glutamic acid); if it were at the 3' end, the last codon would be AAG (arginine), and arginine would be at the carboxyl terminus. Thus, the G is at the 5' end.

7. GUG codes for valine and UGU codes for cysteine. Therefore, an alternating polypeptide containing only valine and cysteine would be made.

8. The repeating tetrapeptide can be read starting with either A, U, A, or G, in that order, yielding the codon sequences, AUA GAU AGA UAG (Ile Asp Arg Stop), UAG AUA GAU AGA (no protein made), AGA UAG (Arg only), and GAU AGA UAG (Asp Arg Stop), if UAG were a stop codon. The second and third sequences would not yield a polypeptide. Detection of either the tripeptide Ile-Asp-Arg or the dipeptide Asp-Arg indicate that UAG is a stop codon. If none of the codons were stop codons, long polypeptides would form. If either AUA, GAU, or AGA were stop codons, different polypeptides would be made

9. (a) The deletion has removed the terminus of one of the coding sequences and the early part of the other coding sequence in a polycistronic mRNA molecule. This is known as a gene fusion. (b) It must be a multiple of three. Otherwise, the carboxyl terminus of the protein would be in a different reading frame and would not match the corresponding terminal sequence of the altered protein.

10. The addition changes the reading frame of the first sequence so that the stop codon for the A protein is not encountered. Thus, the AUG of the B protein is not read. A change producing a stop codon between the start of *A* and the AUG for *B* will stop translation of the *A* gene and allow an in-phase restart at the AUG of *B*. Note that the stop codon can be either before or after the insertion.

11. There is no way of knowing because both codons could be altered to yield the stop codons UAA and UAG.

12. There has been no change in start and stop codons or in coding sequences, so Q is made normally. Note that the situation would be quite different if the inversion not only interchanged *kyu3* and *kyu4* but also interchanged coding and noncoding strands. In that case, neither the *kyu* nor *kyu4* coding strands would be transcribed.

CHAPTER 9   135

13. Mutant #1 could arise in two ways. It could have a deletion of 12 base pairs, those encoding amino acids 1-4 or 2-5. Alternatively, it could have a mutation in the start codon for the first Met. If that were the case, the second AUG would be used for starting. Note that the change in AUG might not work in a prokaryote, because the second AUG might be too far from the Shine-Dalgarno sequence. However, in a eukaryote the first AUG is the one used for starting. Mutation #2 is a change from Tyr (UAU or UAC) to Stop (UAA or UAG); the tripeptide is Met-Leu-His.

14. If the anticodon were ICA, the codon UGA would also be a Cys codon, according to the wobble hypothesis. If GCA were the Cys anticodon, UGA would be reserved as a stop codon. Thus, the Cys anticodon is GCA.

15. Specificity of (1) codon-anticodon binding and (2) recognition of a tRNA molecule by the corresponding aminoacyl synthetase.

16. Protein synthesis cannot start with a 70S ribosome, and soon no 30S particles will be available for forming the preinitiation complex.

17. Each segment of DNA for which there is not a matching RNA appears as a loop in the heteroduplex. Thus, there are seven introns.

18. They will not terminate at the same site because no sequence of four bases contains two overlapping stop codons. They would not usually have the same number of amino acids, but certainly could, depending on where each protein starts and stops.

CHAPTER 10

Mutation and Mutagenesis

CHAPTER SUMMARY

Mutations are abrupt changes in single genes or small regions of a chromosome. When arising in cells that ultimately form gametes, they persist from one generation to the next and may produce phenotypic changes in the progeny. Mutations in other types of cells, somatic mutations, may also produce changes in the organism containing them but are not transmitted to progeny. Most detectable mutations are in genes with protein products and either affect the production of a protein (whether it is produced) or cause an altered protein to be synthesized. Mutations can be classified in a variety of ways--for example, how they come about, the nature of the chemical change, or the way in which the mutation is expressed. Conditional lethal mutations impose changes that are lethal in some (restrictive or nonpermissive) conditions, but not in permissive conditions. Temperature-sensitive mutations are a type of conditional lethal; these cause phenotypic changes or lethality only above or below a particular temperature. Mutations whose origin is unknown are called spontaneous, whereas those resulting from exposure to chemical reagents or physical agents are called induced. A mutation is always a change in the base sequence in a DNA molecule. If a single base is changed, the mutation is a point mutation. If the base change does not affect the amino acid sequence of the protein product or causes an amino acid change that does not affect the activity of the protein, the mutation is said to be silent. A base-substitution mutation may cause chain termination (a nonsense mutation) by production of a stop codon, or an amino acid substitution (a missense mutation) by a change from one amino acid codon to another. Transitions are base-substitution mutations in which the purine-pyrimidine orientation of the base pair is unaltered; if the orientation is reversed, the mutation is a transversion. A mutation may consist of an addition or deletion of one or more bases; if the number of bases is not a multiple of three, a change in the protein product will always occur because of the shift in reading frame of the code, and the mutation is called a frameshift mutation. Large deletions, which may include an entire gene or many genes also occur.

    Spontaneous mutations are random and do not occur in response to environmental conditions; rather, mutants having a growth advantage in certain environments are selected by the conditions and can persist indefinitely. In this way, mutations are the raw material of evolution, with natural selection preserving those mutants which have a selective advantage. The random and nonadaptive nature of mutations has been demonstrated by a statistical test called the fluctuation test and by replica plating. The fluctuation test is also a means of measuring mutation rates with microorganisms. Spontaneous mutations often arise by errors in DNA replication. Such errors occur frequently but are usually corrected by the editing activity of the DNA polymerases and by the mismatch repair system; a small fraction of the errors escape these correction systems and thereby produce a mutation. The editing system removes an incorrectly incorporated base immediately after it is added to the growing end of a DNA strand. The mismatch repair system removes incorrect bases at a later time. Methylation of parental DNA strands and delayed methylation of daughter strands provides the mismatch repair system with information for the direction of correction.

A variety of repair systems exist for repairing chemical and physical damage to DNA. One of these, called SOS repair in bacteria and surely present in other organisms, is error-prone, lacking the editing system, and frequently gives rise to mutations. Error-prone repair is the major cause of mutagenesis by ultraviolet radiation and by alkylating agents. Some types of damage are efficiently repaired by a variety of systems--for example, photoreactivation, excision repair, and recombination repair--and do not lead to mutation. Photoreactivation is a direct cleavage of the pyrimidine dimers produced by ultraviolet radiation. Excision repair is a multistep process in which certain types of altered bases are excised by the combined action of an endonuclease and an exonuclease, followed by repolymerization of the gap produced by the excision. Recombination repair is an exchange process occurring in replicating molecules; gaps in one daughter strand produced by aberrant replication across damaged sites are filled in by nondefective segments from the parental strand of the other branch of the newly replicated DNA.

Mutations can be induced chemically by direct alteration of DNA, for example, by nitrous acid and hydroxylamine. Base analogues, which are molecules able to pair with more than one nucleotide base, are incorporated into DNA during replication, by pairing with a base in the parental strand. In a later round of replication, they pair with other bases, giving rise to transition mutations. 5-Bromouracil is an example of such a mutagen. Acridine molecules interleaf between base pairs of DNA and cause misalignment of parental and daughter strands during DNA replication, giving rise to frameshift mutations, usually of one or two bases. Ionizing radiation causes a variety of alterations in DNA, of which the most obvious are single- and double-strand breaks, which can lead to chromosome breakage, deletions, inversions, and translocations. Less well characterized is the extensive damage of nucleotide bases, though this damage is probably the cause of x-ray-induced point mutations.

Transposable elements are also responsible for many mutations. These elements may interrupt the coding sequence of a gene by insertion within the gene or introduce transcription-termination signals between a promoter and a stop codon.

Organisms having a mutant phenotype sometime revert to the wildtype phenotype. This is called reversion. One mechanism is the restoration of the original wildtype base pair at the mutant site; however, this is not a common event, and reversion is normally a result of an additional mutation at another site. Reversion by secondary mutations is called suppression and the secondary mutations are called suppressor mutations. These can be intragenic or intergenic. In intragenic suppression a mutation in one region of a protein alters the folding of the protein and a change in another amino acid causes correct folding to occur again. Intergenic suppression is of two types, nonsense and missense. In nonsense suppression a chain termination mutation--that is, a stop codon--is read by a mutant tRNA molecule with an anticodon that can interact with the stop codon, and an amino acid is inserted at the site of the stop codon. If the amino acid substitution can be tolerated by the protein in which the original mutation occurred, an active protein will result, and the mutation is suppressed. In missense suppression a mutant tRNA molecule allows insertion of an amino acid other than the one determined by the mutant codon in the mRNA; again, if the amino acid substitution can be tolerated, suppression will occur. Some missense suppression is a result of a mutant aminoacyl synthetases.

The Ames test is a powerful test for mutagenesis by molecules that are not by themselves mutagenic. The Ames test uses histidine-requiring bacteria and measures reversion. A key feature of the reversion test is the use of solid growth medium containing the microsomal fraction of rat liver. Enzymes in this fraction, normally responsible for eliminating toxic substances ingested by mammals, occasionally convert intrinsically harmless molecules to mutagens and dangerous carcinogens. The Ames test is widely used for screening for carcinogens.

BOLD TERMS

acridine, alkylating agent, Ames test, base-substitution mutation, carcinogen, carcinostatic, chain termination mutation, ClB method, conditional lethal mutation, depurination, editing, excision repair, fluctuation test, forward mutation, frameshift mutation, germinal mutation, hot spot, hydroxylamine, incision, induced mutation, intercalation, intergenic suppression, intragenic suppression, ionizing radiation, microsomal fraction, missense mutation, mutation rate, nitrous acid, nonpermissive conditions, nonsense mutation, P element, permissive conditions, photoreactivation, pleiotropic, point mutation, postreplicational repair, pyrimidine dimer, rad, recombination repair, rem, repair synthesis, replica plating, restrictive conditions, reverse mutation, reversion, roentgen, silent mutation, somatic mutation, SOS repair, spontaneous mutation, suppressor mutation, suppressor tRNA, temperature-sensitive mutation, transition, transversion, xeroderma pigmentosum.

ADDITIONAL INFORMATION

The terminology used in discussing mutation is somewhat confusing because of the similarity of four words--mutant, mutation, mutagen, and mutagenesis. These terms have the following meanings. Mutant refers to the genetic state of the organism or the cell. That is, if one of the collection of

characteristics that comprises a so-called normal organism is
different from the "wildtype" character, then that cell or
organism is, in that respect, said to be a mutant. The term
wildtype has been put in quotation marks because it is a
misnomer and often does not refer to the state of the organism
in nature. For example, E. coli isolated from nature is
usually unable to utilize lactose (it is Lac$^-$), yet the
Lac$^+$ phenotype is invariably called wildtype and Lac$^-$ is
called mutant. Generally speaking, the + allele of a gene is
called wildtype and the - allele is called mutant. Mutation
refers to any structural alteration of DNA that is present in
a mutant. A mutation is always a change in the base sequence
of DNA. When a mutation does not cause a phenotypic change, it
is still a mutation, but it is silent. A mutagen is a physical
agent or a chemical reagent that causes mutations to occur.
Mutagenesis is the process of producing a mutation. If it
occurs in nature without the known addition of a mutagen, it
is said to be spontaneous. If a known mutagen is used, the
process is induced mutagenesis. The word mutation is often
used when mutagenesis is meant.

    Four systems for the repair of pyrimidine dimers (T-T, C-C,
and C-T)--photoreactivation, excision repair, recombination
repair, and SOS repair--have been described in the text. The
features of each will be summarized here. These systems differ
both in the source of energy used for repair--blue and
near-ultraviolet light in the case of photoreactivation and
ATP molecules for the other three--and in the mechanisms used
to form functional daughter DNA molecules. Excision repair and
photoreactivation both repair the template, whereas
recombination repair forms a new template. SOS repair instead
ignores the damage and forms uninterrupted daughter strands,
despite the presence of damaged segments of the template. Of
the four, only SOS repair is mutagenic.

    In addition to the modes of repair described in this
chapter a set of enzymes called glycosylases are also
utilized. These fairly recently discovered enzymes cleave the
bond that joins the base to the sugar (the N-glycosylic bond).
Glycosylases are base-specific, for example, removing uracil,
hypoxanthine, or pyrimidine dimers. The uracil glycosylase has
two important functions: (1) it removes uracil that has formed
by spontaneous deamination of cytosine, and it removes uracil
accidentally incorporated as dUTP by DNA polymerases during
DNA replication. When a glycosylase removes a base, the
deoxyribose remains. Such base-free sites are called apurinic
site, since they were first observed as sites of depurination
following treatment of DNA with weak acids or alkylating
agents. An apurinic site can be removed by the following
repair system. First, an enzyme called apurinic endonuclease
makes a single-strand break on the 5' side of the deoxyribose.
The resulting 3'-OH group serves as a primer for DNA
polymerase I in a repair polymerization identical to that used
in excision repair of pyrimidine dimers (see Figure 10-21 of
the text). The next steps are identical to those in excision
repair, namely, displacement of the deoxyribose and a few
additional nucleotides by repair polymerization, excision, and
ligation.

DRILL QUESTIONS

1. A phage which normally produces a large plaque has mutated and makes very tiny plaques on the standard bacterial strain A and normal plaques on another strain B. What class of mutation has altered the plaque morphology?

2. A mutant is isolated that cannot be reverted. What biochemical type(s) of mutation might it carry?

3. Is a change from an AT pair to a GC pair a transition or a transversion?

4. How can tautomerization cause mutation?

5. Will a change in the first base of a codon necessarily produce a mutant protein?

6. Can a somatic mutation be transmitted to progeny?

7. Mutations in what system increase the spontaneous mutation frequency?

8. Some individuals have a patch of white or blond hair in a head of brown hair. What kind of mutation does this represent?

9. Are spontaneous reverse mutations produced randomly. That is, would the fluctuation test give different results for a reverse mutation versus a forward mutation?

10. Could a hydroxylamine-induced mutant be induced to revert by hydroxylamine?

11. Can an acridine-induced mutation be induced to revert by treatment with acridine?

12. Can a mutation induced by nitrous acid be induced to revert by treatment with nitrous acid?

13. A deletion occurs that eliminates a single amino acid in a protein? How may base pairs were deleted?

14. Could a mutation induced by hydroxylamine be induced to revert by treatment with nitrous acid?

15. Which mutagens could cause reversion of a mutation at a hot spot?

16. If a mutagen were added to a culture medium and a fluctuation test was carried out, what, if anything, would be different from the test described in the text?

17. If you knew the base sequence of a wildtype and a mutant, would you known anything about the dominance or recessiveness of the mutation?

CHAPTER 10

18. Name two ways that pyrimidine dimers are removed from DNA.

19. What is the difference between incision and excision?

20. When pyrimidine dimers induce the SOS repair system, are the mutations produced normally at the site of the dimer?

21. Name two ways that a transposable element can prevent expression of a gene.

22. Name two kinds of mutations that would prevent transcription of a gene.

23. A mutation of a bacterial Lac^+ strain yielding a Lac^- colony has been isolated. Several lines of experiments indicate that the mutation resulted from production of a UGA codon. Spontaneous revertants are sought and found at a frequency of 10^{-8} per cell per generation and 9 of 10 of them were caused by suppressor tRNA molecules. What do think is the rate of production of suppressor mutations in the original Lac^+ culture?

ANSWERS TO DRILL QUESTIONS

1. A conditional mutation, probably a suppressor-sensitive mutation?

2. A frameshift or a deletion.

3. A and G are both purines, so it is a transition.

4. A tautomer can cause mutations by having two different base pairing properties that are in equilibrium with one another. A molecule can be incorporated in an aberrant form and then switch to a form that cannot pair with the base in the DNA, or when in the template strand it can temporarily switch its base-pairing properties and allow incorporation of a base that should not be incorporated.

5. No, because an amino acid change does not necessarily produce a mutant protein.

6. No, because it is not in a germline cell.

7. Editing and mismatch repair.

8. A somatic mutation.

9. No.

10. Not if a direct reversal of the mutation was needed. Hydroxylamine causes transitions from GC to AT. A direct reversal would require a reaction with A or T, and hydroxylamine does not react with these bases. However, hydroxylamine could induce a second-site reversion.

11. Yes, because acridines cause both base additions and base deletions.

12. Yes because, nitrous acid causes both AT-to-GC transitions and GC-to-AT transitions.

13. Three. Otherwise, there would have been a frameshift and all downstream bases would have been changed.

14. Yes. The hydroxylamine mutation is the result of a GC-to-AT transition. Nitrous can react with A and cause an AT-to-GC transition.

15. A hot spot is usually a GMeC-to-AT transition. These can be reverted by nitrous acid, which reacts with A. However, if the hot-spot mutation were a deletion, it could not be reverted.

16. The mutation frequency would be higher, but mutations would still occur at random.

17. No. These are phenotypic properties.

18. Photoreactivation and excision repair.

19. Incision refers to the production of one single-strand break, usually near a nucleotide that is to be removed. Excision refers to the second break and subsequent removal of a nucleotide or a set of adjacent nucleotides.

20. No. They can be in any daughter DNA strand that replicated after irradiation.

21. Interrupt a coding sequence; insert a transcription stop signal between the promoter and the stop codon of a coding sequence.

22. Promoter-down mutation, production of a transcription-termination signal between the promoter and the AUG start codon.

23. The mutation frequency in the Lac⁻ culture is 9×10^{-9}. The probability of production of a suppressor is independent of the presence of a suppressor-sensitive mutation (which the Lac⁻ mutation is), so the mutation rate in the Lac⁺ cell is also 9×10^{-9}.

ADDITIONAL PROBLEMS

1. Assume that a single base is chemically altered in a circular phage DNA molecule contained in a phage particle. If the molecule replicates by the θ mode, half of the daughter molecules will retain the parental base sequence and half will have the altered sequence. If the base causes a mutation, half of the progeny DNA molecules will be mutant. Consider a phage whose circular DNA is altered in that way and which replicates

CHAPTER 10   145

exclusively by the rolling circle mode. What will be the distribution of mutants among the progeny of individual bacteria infected with a single phage particle?

2. Since mutation occurs continually in any particular gene, why is it that a population does not become entirely mutant?

3. In microorganisms which mutation rates are more easily measured: ability to inability to synthesize proline (pro^+ to pro^-) or the reverse (pro^- to pro^+)?

4. Assume that in another biological world, also based on DNA and with all of the rules of genetics and molecular biology that we know, the code is nonredundant. How would the mutation rate in this world compare to that in our world? What differences might you expect to find among the subjects discussed so far (through Chapter 10)?

5. Before the genetic code had been elucidated, one proposed code was a code with commas. What was meant by this was that after each codon (whose length was not specified) would be a signal, possible a base sequence, that indicated where the codon stops and where the next one starts. With such a code, what kinds of mutations would not exist?

6. Name three types of mutations that could completely prevent translation of an entire sequence encoding a polypeptide. Which ones would work only in prokaryotes?

7. A mutation is induced in a eukaryotic gene by hydroxylamine. It maps exceedingly close to the beginning of the gene, though its exact location is difficult to establish precisely. When the mutant form of the protein is sought, by looking for a protein with nearly the same amino acid sequence, none is found. You guess that the mutation was not produced by the base analogue, but is instead a spontaneously arising frameshift. To test this idea, you attempt to induce revertants by growth in the presence of acridines, but you are unsuccessful. In fact, revertants can be induced by nitrous acid. Further study of the mutant cells indicates that a completely new protein is present in the cell. It has a slightly smaller number of amino acids than the original wildtype protein, but its amino acid sequence does not resemble the wildtype protein in any way. Other mutations in the original protein do not result in formation of this novel protein. Suggest an origin for this protein.

8. If a culture of bacteria is grown from a phage-sensitive cell, the culture invariably contains phage-resistant mutants. These mutant cells seem to have altered cell walls, since the usual cause of resistance is inability of the phage to adsorb to the cell. In the absence of phage one might expect the fraction of cells that are mutant to increase until the mutation rate equals the reversion rate. This is sometimes the case but for some phages the fraction is much lower in a

minimal medium than in very rich complete media, though the mutation rate is unaffected by the medium. You will not be able to deduce the cause for the phenomenon in any detail, but try to make a general statement about what might be happening. Think about what kind of change might occur in a bacterium that acquires a phage-resistance mutation.

9. Nitrosoguanidine is a potent mutagen in bacteria, so much so that mutations usually occur in clusters. That is, often a mutant bacterium that is selected to have a particular mutation will have acquired other mutations also, and usually these are located near the mutation that is selected. Insight into the mode of action of this mutagen came from studies of bacterial cultures in which growth and cell division were synchronized: all cells started their cycle of growth at nearly the same time and then divided synchronously. When such a cell culture was treated with nitrosoguanidine at various times, it was found that for each time mutations could be detected in only a single region of the map. A plot of mutation frequency as a function of map position and time is shown in Figure 1. What does this suggest about the mechanism of mutagenesis by this compound?

Figure 1

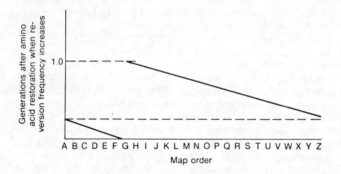

10. Consider a synchronously growing and dividing culture of *E. coli*, as described in Problem 9. In the middle of the replication cycle the cells are given a dose of ultraviolet light. Will the mutations induced by the ultraviolet light be distributed uniformly throughout the map? Explain.

ANSWERS TO ADDITIONAL PROBLEMS

1. In rolling circle replication one strand is a master strand (the one usually drawn in the "center" of the circle). If the base change occurs in that strand, all progeny will be mutant. However, if it occurs in the nonmaster strand (the "outer" strand), only one progeny DNA molecule will be mutant.

2. If the mutation confers a distinct advantage on the organism, it will become prevalent. If it is disadvantageous, it will be lost rapidly. If it is neutral, it will continue to

increase until the forward mutation rate equals the back mutation rate, a phenomenon that is discussed in Chapter 15. Few mutations are totally neutral, though.

3. Ease is of course often a matter of opinion, but in this case there is little doubt that the reverse mutation (pro^- to pro^+) is the more easily detected. Simply by placing bacterial or other microorganisms on two types of solid medium, either with proline or lacking proline, rare pro^+ cells can be detected by their ability to grow on both media.

4. Since 44 of the codons would be nonsense, most base changes would be mutagenic, so the mutation rate would be very high. Silent mutations would be infrequent except in the case of a tolerated amino acid substitution. It is not clear that a genetic system could be able to maintain itself in time with such a rapid mutation rate, since nonviable progeny would often be produced. Thus, it seems likely that highly efficient repair systems would evolve, more efficient than those that exist now.

5. Frameshift mutations, because the correct reading frame would be restored at each comma.

6. (1) A promoter mutation, which would prevent transcription; (2) a mutation in a start codon that would prevent initiation; (3) a mutation in the Shine-Dalgarno sequence, which would prevent binding of the mRNA to a ribosome. The third type could occur only in a prokaryote.

7. A base-substitution near the beginning of a gene should not change the amino acid sequence except for a single amino acid, and the mutant protein would have the same number of amino acids as the wildtype protein. The new protein could have two origins (possibly more): it is made elsewhere in the genome and its synthesis is normally inhibited by the wildtype protein that was mutated; or it is made from the coding sequence of the wildtype protein but in a different reading frame. The first alternative is excluded by the fact that other mutations in the gene do not cause synthesis of the novel protein. The lack of reversion by acridine suggests that a standard type of frameshift in the gene is not present. However, if the mutation was a base substitution and eliminated the AUG start codon and if there was another AUG sequence in a different reading frame, a new polypeptide would be made whose length would be determined by the location of a stop codon in the same novel reading frame.

8. The general statement that can be made is just that phage-resistance mutations must, in some way, be deleterious. The mutant does not have to be lethal, but even if it grew more slowly than wildtype cells, its frequency would be much lower than might be expected. The fact that fewer mutants are present in a minimal medium than in a complete medium suggests

that the mutants have some nutritional problem. This is often the case. Remember that the mutations all affect the cell surface. This might suggest to the more biochemically oriented reader that the mutant cells might have some difficulty transporting some substances from the environment into the cells. This is documented for a few phages. For example, the phage-T1 receptor is actually a transport system for Fe^{2+} ions, which are necessary for the growth of all bacteria; $T1-r$ mutants require a much higher concentration of the ions for rapid growth and in fact tend to grow very slowly in minimal medium unless the concentration of the ion is increased about 50-fold compared to the normal amount. The message you should get from this phenomenon is that mutations are not always what they seem to be and can affect other characteristics of the organism other than the particular one you are examining. This is called pleiotropy.

9. Note that the position of the genes with the maximum mutation rate moves in time along the map. This plus the fact that mutations occur in clusters suggests that nitrosoguanidine acts at the replication fork.

10. Remember that the major cause of mutagenesis by ultraviolet light is the error-prone synthesis of SOS repair. Since this system functions by allowing replication without editing, all mutations must appear in DNA that replicated after the irradiation. The extent to which the mutations will be uniformly distributed depends on the time it takes for the SOS system to be activated and then deactivated. If activation is very rapid and if it does not last for more than one generation time, all mutations will be in the region of the map corresponding to DNA that had not yet replicated during synchronous growth. (This would, by the way, consist of two regions, since $E.\ coli$ DNA is replicated bidirectionally.) If the system is active for more than one generation, mutations will arise throughout the genome. This experiment, to my knowledge, has not been done.

SOLUTIONS TO PROBLEMS IN TEXT

1. $100,000 \times (1 \times 10^{-5}) = 1$ mutation per gamete per generation.

2. $0.5/100,000$, or 5×10^{-6}, since the mutant gene could have come from either parent.

3. (Mutation rate per gene) $\times$ (number of genes per gamete) - (mutation rate per gamete). The minimum estimate of the number of genes is $0.05 \times 10^5 = 5000$, and the maximum estimate is $0.30 \times 10^5 = 30,000$. Note that the minimum estimate is the same as the estimate obtained by counting the number of bands in the polytene salivary gland chromosomes.

4. (a) 5×10^{-5} is the probability of a mutation per generation, so $1 - 5 \times 10^{-5}$ is the probability of no

mutation in one generation. Therefore in 10,000 generations the probability of no mutations is $(1 - 5 \times 10^{-5})^{10,000} = 0.607$ (b) The average number of generations until mutation occurs is $1/(5 \times 10^{-5}) = 20,000$.

5. $\mu = -\ln (5/12)/(5 \times 10^8) = 1.8 \times 10^{-9}$.

6. A: $-\ln (22/40)/(5.6 \times 10^8) = 1.1 \times 10^{-9}$.
   B: $-\ln (15/37)/(5 \times 10^8) = 1.8 \times 10^{-9}$.

7. Proline determines the way the polypeptide backbone folds, so any change from proline (change 1) usually yields a mutant phenotype. Arginine to lysine (change 2) is probably ineffective because both have the same charge and nearly the same size. Substitution of threonine, which is weakly charged, by isoleucine, which is quite hydrophobic (change 3), would very likely cause mutation. Valine to isoleucine (change 4), both of which are nonpolar, would probably have no effect. Glycine to alanine (change 5), both of which are nonpolar, would have little effect unless there was insufficient space for the slightly large alanine side chain. Histidine to tyrosine (change 6) probably would cause a phenotypic change because of the differences in charge and lack of flexibility, though in many proteins the change might be ineffective.

8. Amino acid substitutions at many positions in a protein have little or no effect on its activity; only certain substitutions occurring at critical positions have a detectable effect.

9. (a) No, because hydroxylamine induces CG-to-TA transitions and not TA-to-CG transitions. (b) Yes, since nitrous acid induces transitions in both directions.

10. The glutamic acid codons are GAA and GAG. The lysine codons are AAA and AAG. Therefore, the change in the RNA is G to A, or, in the DNA, a CG-to-TA transition.

11. (a) The glutamic acid codons are GAA and GAG. GAA could be altered to UAA, and GAG could be altered to UAG. (b) If a mutagen that causes GC-to-AT transitions causes reversion, the stop codon must contain a G or a C. Thus, it must be UAG.

12. Six substitutions are possible from UAU and six from UAC. However, because of the redundancy of the code, only six amino acid substitutions in total are possible.

13. The codon changes are (1) Met (AUG) to Leu (UUG, CUG), (2) Met (AUG) to Lys(AAG), (3) Leu (CUX) to Pro (CCX), (4) Pro (CCX) to Thr (ACX), and (5) Thr (AC,A or G)) to Arg (AG,A or G). Note that 1, 2, 4, and 5 are transversion and that 3 is a transition. 5-Bromouracil only induces transitions. Thus, Leu to Pro is the only possible change.

14. It will be incorporated into DNA opposite a T. In a later round of replication it will pair mostly with T but occasionally with C, leading to a GHC pair, at the site of an AT pair. Therefore, the change is an AT-to-GC transition.

15. (a) The four transitions are AT to GC, TA to CG, GC to AT, CG to TA. The eight transversion are AT to cG, AT to TA, TA to GC, TA to AT, CG to AT, CG to GC, GC to TA, and GC to CG. The ratio is one transition : two transversions. (b) $\chi^2 = 0.86$. Agrees, and with one degree of freedom $P = 0.3$--0.5. Therefore, the data agree with the hypothesis.

16. The base change is in one strand of the DNA. Thus, DNA replication yields one daughter DNA molecule with the wildtype gene and another with the mutant gene. Cell division yields one Lac^+ cell and one Lac^- cell. Since the cells do not move on an agar surface, the colony consists of roughly half Lac^+ and half Lac^- cells, which yield purple and pink sectors respectively.

17. In general, the amino acids at a suppressed site are those whose codons differ from the mutant by a single base change. (a) For a UGA codon, the codons are UGC (Cys), UGU (Cys), UGG (Trp), UUA (Leu), UCA (Ser), AGA (Arg), CGA (Arg), and GGA (Gly). (b) For UAA the codons are AAA (Lys), CAA (Gln), GAA (Glu), UUA (Leu), UCA (Ser), UAC (Tyr), and UAU (Tyr).

18. (a) All progeny would be arginine-independent. (b) The suppressor would be present in a 3:1 ratio. These (75 percent of the progeny) will be arginine-independent. (c) Ten percent of the gametes will have an exchange, yielding equal number of arg^-sup^+ and arg^+sup^- gametes. That is, five percent will be arginine-dependent and hence 95 percent of the progeny are expected to be arginine-independent.

19. The two mutations neutralize the effect of one another. The two gene products are probably subunits of a multisubunit protein. The mutant proteins can interact, but a wildtype protein cannot interact with either mutant protein.

20. (a) A double point mutation. (b) Nothing. Frequencies give no information about the type of base change.

21. (a) Lambda apparently does not have a particularly active repair system (in fact, it has none) and is repaired by the bacterial Uvr system. (b) The ultraviolet light has induced the SOS repair system in the bacterium. This system repairs the damage, but it is error-prone in the injected phage DNA. Since the system is error-prone, progeny phage have an increased mutation frequency.

22. Failure to recombine is result of a deletion mutation overlapping either a point mutation or another deletion. A representation of this kind is often called a topological map.

Figure 2

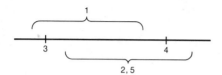

23. Some of the mutagenized P1 particles will have acquired mutations in the a gene. The a^+b^- bacterial strain is infected with the mutagenized P1 population and b^+ transductants are selected. These are tested for the a allele. Some of the transductants will be a^- because of linkage between the a and $5b$ genes.

CHAPTER 11

Regulation of Gene Activity

CHAPTER SUMMARY

Most cells do not synthesize molecules that are not needed. From the point of view of cellular economy, proteins may be subdivided into three classes: those required continuously, those required only in certain environments, and those whose concentration should, for the sake of efficiency, vary with the concentration of certain substances in the environment. How the synthesis of proteins of each class is regulated differs between prokaryotes and eukaryotes.

In bacteria the synthesis of most proteins is regulated by controlling the rate of transcription of the genes encoding the proteins. Enzymes needed continuously, such as those required to metabolize glucose, must be made continuously; however, to avoid excessive and wasteful synthesis of the enzymes, the amount of transcription is often regulated by the concentration of the enzyme itself. A variety of mechanisms can accomplish this type of regulation, which is called autoregulation, but the most common one is direct binding of the enzyme to a base sequence in the DNA near the promoter for the gene. When the sequence is occupied, transcription does not occur. Thus, if the concentration of the enzyme were to fluctuate for any reason, the rate of transcription would also fluctuate, but in a way that would restore the enzyme concentration to the required value.

The synthesis of degradative enzymes needed only on occasion, such as the enzymes required to metabolize lactose, a sugar encountered only infrequently in the environment, are regulated by an off-on mechanism. When lactose is present, transcription of the genes encoding the enzymes required to metabolize lactose are made; when lactose is absent, such transcription does not occur. Lactose metabolism is regulated by a mechanism called negative regulation. Two enzymes required to degrade lactose--permease (required for entry of lactose into bacteria) and beta-galactosidase (the actual degrading enzyme)--are encoded in a single polycistronic mRNA molecule, *lac* mRNA. Immediately adjacent to the promoter for *lac* mRNA is a regulatory sequence of bases called an operator. A repressor protein is made by still another adjacent gene, and this protein binds tightly to the operator, thereby preventing RNA polymerase from initiating transcription at the promoter. Lactose is an inducer of transcription, because it can bind to the repressor, thereby preventing the repressor from interacting with the operator. Therefore, in the presence of lactose, there is no active repressor, and the *lac* promoter is always available to RNA polymerase. The repressor gene, the operator, the promoter, and the structural genes are adjacent to one another (separated only by small spacers, except between the promoter and the operator); together they constitute the *lac* operon. Repressor mutations have been isolated that inactivate the repressor protein, and operator mutations are known that prevent recognition of the operator by an active repressor; such mutations cause continuous production of *lac* mRNA and are said to be constitutive.

When lactose is cleaved by beta-galactosidase, the products are glucose and galactose. Glucose is metabolized by enzymes that are made continually; galactose is broken down by another inducible operon, the galactose operon. When glucose is present in a growth medium, the enzymes for degrading lactose are unnecessary, so the following general mechanism for preventing transcription of many sugar-degrading operons has evolved in bacteria. High concentrations of glucose suppress the synthesis of a small molecule--cyclic AMP (cAMP). Initiation of transcription of many sugar operons requires binding of a particular protein molecule, called CAP, to a specific region of the promoter for the operon. Binding occurs only after CAP has first bound cAMP and formed a cAMP-CAP complex. Only when glucose is absent is the concentration of cAMP sufficient to produce cAMP-CAP and hence to permit transcription of the sugar operons. Thus, in contrast with a repressor, which must be removed before transcription can begin (negative regulation), cAMP is a positive regulator of transcription.

Biosynthetic enzymatic systems exemplify the third class of protein-synthesizing system. Such a system operates most efficiently when the concentration of each component is determined by the amount of the reaction product in the growth medium. For example, in the synthesis of tryptophan, transcription of the genes encoding the *trp* enzymes is controlled by the concentration of tryptophan in the growth medium. The resulting rate of enzyme synthesis is always sufficient to maintain a rate of tryptophan synthesis necessary for adequate overall protein synthesis. Amino acid-synthesizing operons are usually regulated by attenuator systems. Transcription is initiated continually but terminated at a site ahead of the genes encoding the enzymes. The frequency of termination of transcription is determined by the availability of tryptophan; with decreasing concentration of tryptophan, termination occurs less often and the tryptophan-synthesizing enzymes are made, thereby increasing the concentration of tryptophan.

Regulation of genetic systems in eukaryotes is accomplished in a variety of ways, most of which are quite different from the mechanisms observed in eukaryotes. In higher organisms the environment of the cells is fairly constant, and cells are instead usually called upon to respond to signals coming from within the organism. The response may be a temporary one, such as synthesis of a digestive enzyme, or a permanent change to a differentiated cell. Eukaryotic genes are rarely arranged in operons consisting of adjacent genes, but form more loosely organized systems called gene families. Genes in a family are usually not adjacent and may even be located on different chromosomes. Several genes in a family can often be activated simultaneously; it is assumed that each gene has a common regulatory region that responds to a particular signal, though signals of this type have not yet been identified. Some gene families consist of multiple copies of a single gene.

Differentiation and production of particular eukaryotic

gene products (either protein or RNA) are induced in several
different ways, the method depending on the requirements of
the cells. For example, the enormous amount of ribosomal RNA
needed during development of an amphibian egg is obtained by
specific replication of the rRNA genes, thereby producing a
huge number of these genes, each of which can be transcribed;
this phenomenon is called gene amplification. In contrast, the
production of silk fibroin during cocoon formation by moths is
accomplished by synthesis of a long-lived mRNA molecule, so
that translation can continue for several days. Transcription
is also turned on and off in numerous eukaryotic systems,
particularly those controlled by hormones. Presumably, either
a macromolecular complex containing the hormone binds to a
site on the chromosome and acts as a positive regulator, or
the hormone complex removes a negative regulator; while these
ideas seem reasonable, direct experimental confirmation is not
yet available for many systems.

Differential processing of primary RNA transcripts is
also used in regulating synthesis of particular proteins. For
example, in different cell types unique patterns of splicing
may produce two mRNA molecules having distinct leader
sequences for a particular gene, resulting in different
efficiencies of translation in each cell type. Different
patterns of splicing also have been observed to produce
different proteins from the same primary transcript.
Regulation of the site of addition of poly(A) termini can also
yield different mRNA molecules from a single primary
transcript.

Base sequences upstream from some eukaryotic promoters are
required for initiation of transcription. Those known as
upstream activation sites may be hundreds of nucleotide pairs
away from the promoter, a remarkably large distance. The
events that occur at such sites are not known. Enhancer
sequences also increase the rate of transcription when located
near most eukaryotic genes. Remarkably, enhancers are equally
effective in both forward and reverse orientations. The
mechanism of enhancement is unknown. Certain regions of
chromatin upstream from particular genes change structure when
the genes are expressed. In isolated chromatin these regions,
known as hypersensitive sites, are excessively sensitive to
degradation by nucleases; it has been suggested that these
regions may lack nucleosome structure. Possibly,
hypersensitive sites are equivalent to upstream activation
sites.

In prokaryotes the relative concentrations of
functionally related proteins are usually maintained by the
encoding of each protein in a single polycistronic mRNA
molecule. Different efficiencies of translation of each gene
determine the rate of synthesis of each protein. However,
eukaryotic mRNA is not polycistronic, as only a single protein
can be translated from a eukaryotic mRNA molecule. In
eukaryotes, equal concentrations of related proteins are
frequently obtained by synthesis of a long polypeptide called
a polyprotein. The amino acid sequences of several proteins
may be contained in the polyprotein; hence, cleavage of the
polyprotein at particular sites yields distinct proteins. The

individual proteins are usually present in equimolar
quantities, since one copy of each amino acid sequence is
typically contained in the polyprotein.
   There is no universal principle of regulation of gene
expression, either in prokaryotes or eukaryotes. Evolution has
probably merely selected a mechanism that works for each
situation.
   The activity of enzymes is also frequently regulated, in
order to avoid wasted consumption of energy and precursor
molecules. In an unbranched biochemical pathway, commonly the
product of the reaction sequence inhibits the first enzyme in
the pathway. This inhibition is called end-product or feedback
inhibition. In a branched pathway the end product of each
branch usually inhibits the first enzyme following the branch.
Occasionally, in branched pathways isoenzymes (enzymes that
catalyze the same reaction but respond to different
inhibitors) catalyze the first step in the pathway. When this
occurs, the end product of each branch usually inhibits one of
the isoenzymes, so that intermediates in the pathway are
synthesized in quantities appropriate for synthesis of the
products that are not present.

BOLD TERMS

adenyl cyclase, allosteric effector molecule, antitermination,
aporepressor, attenuator, autoregulation, beta-galactosidase,
branched pathway, cAMP, cAMP-CAP, CAP protein, catabolite
activator protein, *cis*-dominant, complex multigene family,
constitutive, coordinate regulation, cyclic AMP, derepression,
developmentally controlled gene family, enhancer element,
estrogen, feedback inhibition, gene amplification, gene
dosage, gene family, gene regulation, hormone, hypersensitive
site, inducer, inducible, isoenzyme, *lac* operon, *lac*
repressor, lactose permease, leader, leader polypeptide,
masked mRNA, negative regulation, operator, operon model,
polarity, polyprotein, positive regulation, pseudogene,
recruitment factor, simple multigene family, translational
regulation, upstream activation site.

ADDITIONAL INFORMATION

Experience in teaching has indicated that one of the most
difficult things for a student learning about regulation is
the determination of the expression of an operon in a partial
diploid when there are *cis*-dominant mutations. This is an
important skill, since analysis of partial diploids is a
standard way to distinguish positive and negative regulation
and to distinguish structural genes from sites in the DNA.
Thus, as a learning aid, a few examples will be worked out for
the *lac* operon.
   Example I: $lacZ^+ lacY^- / lacZ^- lacY^+$.
In this simple partial diploid there are no *cis*-dominant
mutations. Note that the location of the two genotypes (that
is, is to the left or the right of the slash) is unimportant,
since expression from a chromosome or a plasmid is identical.
We ask whether beta-galactosidase is made and, if so, whether

its synthesis is constitutive or inducible. The genotype at the left is capable of making the enzyme, whereas that on the right is not; nonetheless, the cell will contain enzyme, as long as the operon can be turned on. To turn it on, functional regulatory elements need to be present, as they are, and an inducer must enter the cell. The latter can be done because the genotype at the right can be induced to make the permease. Thus, for this partial diploid beta-galactosidase can be made, and its synthesis is inducible.

Example II:
$lac^C lacZ^- lacY^+ / lacZ^+ lacY^-$.

This partial diploid has the *cis*-dominant mutation lac^C. Note that the genotype at the left contains the constitutive mutation lac^C, so the genes at the left are expressed. However, from the *lacZ* gene at the left a defective enzyme is made. At the right is a functional *lacZ* gene from which active enzyme can be made. However, its synthesis is under the control of a normal operator (note that the operator genotype is not indicated, so by convention it is $lacO^+$) and hence enzyme synthesis must be induced. Thus, the partial diploid makes a defective enzyme constitutively and a normal enzyme by induction, so all together the cell can make the enzyme by induction.

Example III:
$lacP^- lacZ^+ / lac^C lacZ^-$.

At the left is a promoter mutation, which is *cis*-dominant. This means that no *lac* mRNA can be made from this segment of DNA, and the genes may be thought of as being absent. That is, the genotype of the cell can be considered to be just that at the right, which makes a defective enzyme constitutively, as in Example II. Thus, there is no way for the cell to make active enzyme.

Example IV:
$lacP^- lacI^+ lacZ^+ / lacI^- lacZ^+$.

The genotype at the left contains a promoter mutation, so the *lacZ* and *lacY* genes can be considered absent. However, the *lacI* gene has its own promoter, so *lac* repressor molecules will be present in the cell. The genotype at the right could alone make enzyme constitutively because of the *lacI* mutations. However, a functional *lacI* product is made by the genotype at the left. Thus, the cell can make enzyme, and synthesis must be induced.

Example V:
$lacP^- lacI^+ lacY^+ / lacI^- lacY^-$.

This genotype differs from that in Example IV by the presence of a $lacY^-$ mutation on the right. Again, the genotype at the right contributes only *lacI* repressor to the cell, so any synthesis of the enzyme must be inducible. However, the genotype at the right is $lacY^-$, indicating that inducer cannot enter the cell. Thus, this cell is unable to make any enzyme, because synthesis would have to be induced and there is no way to induce it.

The *lac* operon is considered to be a prototype operon. However, the student must not infer that all operons for degradative systems are controlled like the *lac* operon. For example, the operon responsible for degradation of the sugar arabinose utilizes both a positive and a negative

regulator. Space is not available for describing this complex
system, but suffice it to say that the inducer, which is
arabinose, binds to the AraC protein, converting it to a form
that can bind to DNA near the *ara* promoter and thereby
initiation transcription. Details of this system can be found
in Chapter 14 of Freifelder, *Molecular Biology*.
   The operon responsible for degradation of galactose (the
gal operon) is interesting, because it has two promoters
for the same genes, and these promoters are adjacent. This
operon is negatively regulated by a repressor and positively
regulated by cAMP-CAP, as with the *lac* operon. However,
only one of these promoters, the stronger one, is regulated by
cAMP-CAP. Galactose serves two functions in *E. coli*. One
function is as a carbon source when it is present and glucose
is absent, exactly as with lactose. In this role, the strong
promoter should be subject to repression and to regulation by
cAMP-CAP, for the same reasons as with the *lac* operon.
However, one of the *gal* enzymes is also required for
synthesis of a component needed to form the cell wall. If
synthesis of this enzyme were inhibited by glucose, cell wall
formation would stop; thus, this enzyme must be made
constitutively. This need by the cell could be satisfied by
having two independent sets of genes, one solely for
metabolizing galactose when glucose is absent (this one would
be subject to repression and cAMP-CAP regulation) and another
that is constitutive. However, evolution has chosen not to
duplicate one of the genes but to duplicate the promoter,
providing a strong promoter that can be shut down and a weak
promoter that is always available.
   A great deal is known about regulation in both
prokaryotes and in eukaryotes and only a sampling has been
presented in *General Genetics*. The student interested in
learning more about regulation in eukaryotes should consult
the book *Genes,* by B. Lewin (1983, Wiley).

DRILL QUESTIONS

1. Which type of regulation, positive or negative, involves removal of an inhibitor?

2. Would synthesis of an enzyme that is needed continually be regulated?

3. Which enzymes of the *lac* operon are regulated by the repressor?

4. What is meant by a partial diploid?

5. Physically what is the consequence of binding of the *lac* repressor to the *lac* operator?

6. What term describes a gene that is expressed continually, even though its transcription may be autoregulated?

7. Is the *lac* repressor itself made constitutively or is it induced?

8. What is the biochemical action of an inducer?

9. Does an inducer necessarily inactivate a repressor?

10. Is an inducer necessarily an inactivating agent?

11. When glucose is present, is the concentration of cyclic AMP high or low?

12. Can a mutant with either an inactive adenyl cyclase gene or an inactive *crp* gene synthesize beta-galactosidase?

13. Does the binding of cAMP-CAP to DNA affect the binding of a repressor in any way?

14. Coordinate regulation is a way to turn on and off the synthesis of a collection of enzymes having related function. What else is accomplished by having such gene organization?

15. Are all proteins translated from a single polycistronic mRNA necessarily made in the same quantity?

16. Why are all constitutively synthesized proteins not made at the same rate?

17. Repressors and aporepressors both bind molecules that are components of the metabolic pathway encoded in an operon. What is different about the positions in the pathway of the molecules bound, and what is different about the activity of the complex formed when binding occurs?

18. Is the attenuator, like the operator, a binding site?

19. Is RNA synthesis ever initiated at an attenuator?

20. Antitermination and attenuation are both concerned with termination of transcription. How do they differ, in principle, with respect to the role of RNA polymerase?

21. In feedback inhibition of a single unbranched pathway, which substance is the inhibitor and which enzyme is inhibited?

22. In a feedback-inhibited pathway with two branches, if isoenzymes were utilized in regulation the activity of the pathway, how many isoenzymes would normally be used?

23. Are the individual members of a gene family necessarily coordinately regulated?

24. Name two mechanisms utilized in eukaryotes for maintaining a fixed ratio of gene products, when equality of the products is not required.

25. With reference to Question 24, what mechanism are utilized

when equality of the gene products is required? Give two examples.

26. In a eukaryote an extracellular regulator forms an intracellular complex that ultimately binds to DNA. What barriers must it pass in order to reach the DNA?

27. What biochemical process is usually regulated by hormones that control synthesis of a particular gene product?

28. Do hormones ever bind directly to DNA?

29. Most regulatory elements are in fixed positions with respect to the gene(s) regulated. Which element is not?

30. If an enhancer were moved from a position near gene A, where it has strong enhancing activity, to a position 50 nucleotides upstream from gene B, which is constitutively transcribed, would transcription of gene B be increased?

ANSWERS TO DRILL QUESTIONS

1. Negative regulation.

2. Not always, especially if needed in large quantities. If it were regulated, it would usually be autoregulated.

3. Beta-galactosidase (*lacZ* product), lactose permease (*lacY* product), and the transacetylase (*lacA* product).

4. A cell in which two copies of a small number of genes are present.

5. The promoter is inaccessible to RNA polymerase.

6. It is constitutive.

7. Constitutively.

8. It binds to a repressor.

9. Yes, by the definition of a repressor.

10. No. In a negatively regulated system it is an inactivator. In a positively regulated system it is an activator.

11. Low.

12. No. In the first case, it could not make cAMP. In the second case, it could not make CAP.

13. No.

14. The ratio of the amount of each components is maintained constant without having each one separately regulated.

15. No. There is often a polar effect in that the relative amounts of each protein decrease toward the 3' end of the mRNA.

16. The major reason is that all promoters do not have the same strength, so the rate of initiation of transcription varies from one gene to the next.

17. With a repressor the molecule is usually an early (generally the first) reactant in the pathway. The repressor is inactivated by the binding. With an aporepressor the molecule is usually the product of the pathway. An aporepressor is activated by the binding.

18. No. Nothing is specifically bound to an attenuator.

19. It is strictly a potential termination site for transcription.

20. In attenuation the structure of the termination sequence is altered so that RNA polymerase either does or does not see it. In antitermination RNA polymerase is altered so that it can ignore certain termination sites.

21. The product of the pathway is usually the inhibitor, and the first enzyme in the pathway is usually inhibited.

22. Two.

23. No. A gene family may consist of related genes that are expressed fairly independently of one another--for example, the developmentally regulated families.

24. (1) Differential processing of a single primary transcript. (2) Common regulation of the synthesis of different transcripts.

25. The general mechanism is to break down a precursor into subunits, each of which is one of the desired products. One example is the cleavage of an RNA precursor to obtain rRNA components. A second example is cleavage of a polyprotein.

26. First, it must pass through the external cell. Then, it must pass through the nuclear membrane. Finally, it may have to disrupt nucleosome structure and remove histones, before it can actually reach the DNA.

27. Transcription.

28. No examples are known and it is quite unlikely. The active form is probably a hormone-receptor complex, which may bind directly to DNA or may activate another binding protein.

29. Enhancer sequences.

30. Very likely, unless gene B was somehow separated from some element needed for its activity.

ADDITIONAL PROBLEMS

1. It is known that the initiation of DNA synthesis in prokaryotes is regulated, because inhibition of protein synthesis prevents initiation. Can you suggest any ways by which synthesis of an initiation protein might be regulated?

2. Is it fair to say that a constitutive gene is unregulated?

3. Is it necessary in a bacterial operon for the gene for a repressor to be near the structural genes?

4. Consider an attenuated operon in which a regulatory protein could bind to the attenuation sequence and enable RNA polymerase to ignore the attenuator. The regulatory protein is activated by the substrate of one of the enzymes of the operon. Is this an example of positive or negative regulation of the operon?

5. An operon has the gene sequence A B C D E. Neither the promoter nor the operator have been located. The repressor gene maps very far away from the structural genes. Various deletion mutants have been isolated. Some deletions of gene E, but none for any of the other genes, result in constitutive production of the mRNA of the operon. Where do you think the operator and the promoter are?

6. If an operon is derepressed and then, at a later time, repressed again, will the ability to carry out the enzymatic reactions determined by the structural genes of the operon be eliminated as rapidly as transcription is turned off?

7. This is a slightly complex problem that requires thinking about both the result of Additional Problem 6 and some features of the biology of phage lambda. Lysogenization of phage lambda is controlled by a repressor encoded in a gene cI. In a lysogen if the repressor is inactivated by any of several means, lytic development of the phage ensues, with the prophage being excised from the bacterial chromosome. A useful repressor mutant is called cI857. It is temperature-sensitive, and merely maintaining the temperature of a culture of a lysogen at 42 C causes inactivation of the repressor and subsequent phage development. An interesting feature of the temperature sensitivity is that it is reversible. That is, if a culture is heated to 42 C for two minutes and then returned to 30 C, phage are not formed and the lysogen remains stable. On the other hand, if the culture is heated for 20 minutes and then cooled, progeny phage are made and the bacteria lyse. What do you think is the difference between the 2-minute heating and the 20-minute heating?

8. Referring again to Problem 7, a culture is heated to 42 C for 5 minutes, cooled and grown for about one hour during

which time the cells divide once or twice. The culture is then restored to 42 C and maintained at that temperature until phage develop and the cells lyse. More than half of the bacteria do not produce phage and are able to form colonies when plated on a solid medium at 42 C. Explain this phenomenon.

ANSWERS TO ADDITIONAL PROBLEMS

1. This problem has been widely discussed and the answer is not known. However, two ideas are the following: (1) Replication of a gene that is one of the last genes replicated causes transcription, and the gene product is an initiator protein. (2) Since cells grow continually and replication is coordinated with growth, the concentration of an inhibitor continues to drop as the cell enlarges. The inhibitor is thought to be a multisubunit protein in equilibrium with the monomers. When the cell volume exceeds a certain point, dissociation of the multimer causes the concentration of the inhibitor to drop below a critical value, and inhibition of initiation is relieved.

2. No. Constitutivity simply means that transcription of the gene is turned on all of the time. The rate of transcription may be modulated by autoregulation, processing of the RNA may be regulated, and there may be some mode of translational control.

3. No. The repressor is a diffusible protein, so the gene be anywhere. In fact, there are many repressors, of which the *trp* repressor is one example, whose genes are located quite far from the structural genes. The only reason that a repressor gene is frequently near the structural genes is that in the course of evolution they will tend to approach one another to reduce the probability of being separated by recombination, translocation, and inversion.

4. It is negative regulation. The attenuator prevents transcription of the coding sequence of the gene, and the attenuator is inactivated by the regulatory protein.

5. The operator is probably next to gene *E*. Deletions that remove the operator make transcription constitutive. The promoter cannot be between the operator and gene *E*, because in that case all deletions that remove the operator would also remove the promoter, and no constitutive deletions would be found. The order is *E o p*, which is consistent with the fact that some deletions are not constitutive; these would be larger deletions that include both the operator and the promoter. Thus, the gene order (transcription from left to right) is *p o E D C B A*.

6. No. There will be no further synthesis of the enzymes but the enzymes will persist. In time (but a long time) the enzymes might be degraded. If the cell continues to divide the enzyme activity will gradually decrease, owing to distribution

of the enzyme molecules among daughter cells.

7. After 2 minutes insufficient mRNA has been made. By 20 minutes so much has been made and translated that preventing further synthesis of mRNA cannot stop phage production.

8. Cells that can form colonies at 42 C cannot contain a prophage with the *cI857* mutation. They are nonlysogens, which means that excision must have occurred. Since the cells did not lyse or produce phage during the hour at the lower temperature, repression must have been reestablished. Clearly, during the 5-minute heating enzymes needed for phage development were not synthesized. Since excision has occurred, the integrase and excision system must have been made. Thus, during the low-temperature period, prophage excision occurred, but without replication or further development. Thus, when the cell divides, the excised prophage is present in one daughter cell and not the other. When a cell with an excised prophage is reheated, mRNA is made and normal phage development occurs. The daughter cell without the excised prophage contains no lambda DNA and hence is unaffected by further heating.

SOLUTIONS TO PROBLEMS IN TEXT

1. Repressed: transcription inhibited; induced: repressor inactivated; constitutive: never repressed; coordinate regulation: control of synthesis of a set of proteins encoded in a single polycistronic mRNA by a single control element. See also the Glossary of *General Genetics*.

2. Since it is needed continuously, it need not be repressed by substrates or positive effectors. However, the amount of all proteins should be regulated so that there is neither runaway nor inadequate synthesis. Systems of this sort are often autoregulated in that the rate of synthesis of the enzymes depends on the amount of enzyme present. Thus, as a cell grows and enlarges, the concentration of the enzymes drops and more enzyme is made in order to maintain a constant concentration.

3. If neither glucose nor lactose are present, two proteins (the repressor and CAP-cAMP) are bound to the DNA. If glucose alone is present, only one protein (the repressor) is bound to the DNA.

4. It seems likely that on occasion, perhaps once per cell generation, a *lac* transcript will be made (that is, no system is ever really completely turned off), so there will be a very small number of permease molecules in the cell. These few molecules could allow the first lactose molecule to enter the cell.

5. First, *lac* mRNA and, then, beta-galactosidase is made. When the lactose is finally consumed, synthesis of *lac* mRNA will cease, and no more enzyme will be made. The enzyme already made will persist, so the enzymatic activity remains;

however, the enzyme concentration drops as it is diluted out by cell division.

6. The *lac* genes are transferred by the Hfr cell and enter the recipient. No repressor is present in the recipient, so *lac* mRNA is made. However, a *lacI* gene also is transferred to the recipient, so soon afterward repressor will be made and *lac* transcription will stop.

7. The cell cannot be derepressed. The simplest explanation is that the repressor cannot bind the inducer. The partial diploid given at the end of the problem would not make enzyme, because the constitutive component makes no enzyme and the other component cannot be induced.

8. (a) When X is lacking from the medium, Xase is without value. Cells in which 20 percent of their protein is useless must grow slightly more slowly. If the culture medium does not limit growth for other reasons, growth will be limited by how fast the necessary quantities of the required proteins can be made. The repressorless constitutive strain should grow 20 percent slower; that 20 percent disadvantage will be compounded in each generation. When the inducible strain has undergone 30 doublings, the constitutive one will have undergone only $0.8 \times 30 = 24$ doublings. At that time, for every 2^{30} cells of the inducible strain, there are only 2^{24} cells of the constitutive strains--that is, 2^6 times as many cells of the former as of the latter. Therefore, the ratio of wildtype (inducible) cells to repressorless (constitutive) cells is $2^6 = 64$. (b) If a preferred carbon source is also present, the answer is essentially the same as in (a). If there is no other carbon source but X, the constitutive strain is no longer at a disadvantage. (c) An inducible mutant, owing to a repressor that is defective in binding the gratuitous inducer. The reasoning is the following. The gratuitous inducer is not a substrate of Xase; that is, it is not a carbon source. Ability to remain induced would confer a growth advantage on a cell in these special conditions, so if such a mutation arose spontaneously, it would be selected for in these circumstances.

9. (a) The strain is constitutive, that is, *lacI*$^-$ or *lac*^{c}. (b) The second mutant must be *lacZ*$^-$ since no beta-galactosidase is made; also, it must have a normal repressor-operator system, since permease synthesis responds to lactose. That is, its genotype is *lacI*$^+$*lacO*$^+$*lacZ*$^-$*lacY*$^+$. Since the partial diploid is regulated, the operator in the operon with a functional *lacZ* gene must be wildtype. Hence, the first mutant must be *lacI*$^-$.

10. It clearly lacks a general system that regulates sugar metabolism. The only thing we know about is the cAMP-dependent system, namely, that several mutations prevent activity of this system. For example, the mutant could make (1) either a defective CAP protein, and either fail to bind to the promoter

or be unresponsive to cAMP, or (2) an inactive adenyl cyclase gene, making no cAMP.

11. Since addition of Q causes Qase to be made, the system is regulated. Deletion of *kyu* prevents induction, suggesting that *kyu* encodes a positive regulatory element. A partial diploid with both wildtype *kyu* and the deletion is regulated, indicating that the *kyu* product is present and that it is a positive regulator activated by an inducer. Examination of mutants confirms this view. The *kyuI* mutant never makes Qase but a partial diploid with wildtype is inducible, indicating that the *kyuI* mutation is recessive. It seems likely that the *kyu* product binds Q and is activated. The *kyuII* mutation is constitutive and dominant, suggesting that it is able to turn on transcription of Qase without Q. The *kyuI* product probably fails to bind Q and the active positive regulator is a Q-Kyu complex. The *kyuII* product is probably able to bind to the promoter without first binding Q.

12. Glucose is irrelevant to the functioning of the *trp* operon. When tryptophan is present (a and c), the repressor is active and is bound to the DNA.

13. (a) No, 2. If the promoter cannot bind RNA polymerase, the adjacent gene cannot be transcribed. (b) Yes, 2. If the operator cannot bind the repressor, there will be no repression and transcription of adjacent genes cannot be shut off. (c) Yes, 3. The repressor cannot bind to the operator, so the system is always on; however, in a partial diploid in which a nondefective repressor is made, this repressor can bind to the operator and prevent transcription. (d) No, 1. The repressor cannot be inactivated. It will bind to all *his* operators and prevent all *his* transcription. (e) No, 1. The tRNA cannot be prevented from activating the repressor. Therefore, all *his* operators will be occupied and all *his* transcription will be prevented.

14. (a) G inhibits 5; H inhibits 7; G and H together or E alone probably inhibit 3; J inhibits 8; enzyme 1 could be inhibited by (G, H, and J), (E and J), or (C and E); ( ) enclose the products that act together. (b) Step 1, three isoenzymes; step 3, two isoenzymes.

15. (a) Eukaryotic genes are not generally organized into polycistronic operons, because without complex splicing events, only one protein molecule can be translated from a primary transcript. (b) An operon is a set of coordinately regulated genes encoded in one or two polycistronic mRNA molecules. If there is a regulatory gene (for example, a repressor), it is usually encoded in a separate monocistronic mRNA molecule. A gene family is a collection of genes, each of which yields a distinct RNA molecule, which encode molecules of similar or related function. The genes may or may not be clustered and they are rarely adjacent, but they are somehow related by similar or related signals.

16. No, for all three parts. This conclusion cannot be rationalized; it is simply an experimental fact.

17. Gene amplification is used when a limited time is available, because by increasing the number of genes, the rate of production of a particular protein per cell can be increased greatly. When a long time is available, achieving a high rate of synthesis is not necessary; instead, continued synthesis of mRNA and extended lifetime of mRNA accomplish the end.

18. In gene amplification the number of copies of a gene is increased; therefore, more mRNA can be made per unit time. In translational amplification, the lifetime of the mRNA is increased, which enables a great deal of mRNA to be present at a particular time.

19. (a) No, because there is no DNA to be transcribed. (b) This is reasonable, because no other protein is being made by these cells. In fact, it is true and the regulator is heme.

20. The specificity lies in the chromatin not in the extract and some inhibitor of oocyte transcription is washed from the chromatin by the 0.6 M NaCl.

21. The hormone causes a primary transcript to become smaller and to be able to make a functional gene product. Thus, the hormone is apparently needed to initiate intron excision.

22. (a) The promoters for both transcription units probably have a common sequence acted on by either a positive or negative regulator. (b) Both primary transcripts have a common sequence acted on by an element that prevents some stage of processing. (c) Both processed mRNA molecules have a common sequence involved in ribosome binding. An effector may remove a protein bound to this sequence or it may denature a double-stranded region containing the ribosome binding site.

23. (a) Yes, I, yes. Both left and right sides contain diffusible elements, so complementation occurs. (b) Yes, I, yes. The right side is constitutive for synthesis of a defective enzyme. Thus, the enzyme is made from the left side, which is inducible. (c) Yes, C, yes. The lack of the repressor on the right side is irrelevant. The right side makes enzyme constitutively. (d) Yes, I, yes. Everything complements perfectly here. (e) Yes, C, yes. Both z and y products are made. There is no active repressor in the cell, so synthesis is constitutive. (f) Yes, I, yes. The right side makes an inactive enzyme constitutively, which is of no value. However, it also makes the repressor, which acts on the left side to make synthesis of enzyme inducible. (g) No, neither, no. The left side makes no enzyme because of the the promoter mutation. The right side can only make a defective enzyme. (h) Yes, I, yes. The left side makes a defective enzyme constitutively, which is of no value. The active y product

compensates for the inactive *y* product made by the right side. The right side makes enzymes but only after induction. (i) No, neither, no. The operator mutation at the left contributes nothing, since the promoter mutation prevents expression of the *z* and *y* genes. The right side is, in principle, inducible; however, because of the *y* mutation, no inducer can get into the cell. The left side also makes no permease because of the promoter mutation. (j) Yes, neither, no. There is no repressor made in the cell, so the enzyme is made constitutively from the right side. However, the right side cannot make active permease, and the left side makes no permease. Thus, the enzyme is made but the cell cannot grow on lactose, because lactose cannot enter the cell.

24. (a) The small plasmid which carries the *lac* operator sequence is present to the extent of 20 to 30 copies per cell. There are few molecules of *lac* repressor in each cell. The 20 to 30 "extra" operator sequences that are not situated in a *lac* operon compete with the chromosomal *lac* operator sequence for the binding of *lac* repressor, resulting in at least partial induction. (b) A "promoter-up" mutation in the *lacI* gene which generated overproduction of *lac* repressor could produce such a result.

25. (a) The PhoR protein, because in its absence the expression is insensitive to the presence or absence of phosphate. (b) The PhoB protein; it is a positive effector. The PhoB protein must be binding to DNA and acting as a positive effector necessary for synthesis of APase. If the PhoR protein were acting directly on the DNA, then it would have to be a repressor. Mutants of the *phoR* gene do show constitutivity, as repressor mutants do, but a *phoB⁻phoR⁻* double mutant would also show constitutivity if the *phoR* gene were a direct repressor, and they do not. Therefore, since the PhoR protein cannot be a direct repressor of transcription, the PhoB protein must be a direct positive control element affecting transcription of the *phoA* gene. (c) If the PhoB protein binds to the DNA at a promoter-like spot, and the PhoR protein in the presence of phosphate acts to reduce the synthesis of the PhoA protein, then the PhoR protein probably forms a complex with the PhoB protein and phosphate, making it harder for the PhoB protein to bind at the promoter site. Thus, transcription is inhibited. The *phoB⁻* mutation allows only five percent expression of the *phoA* gene in the absence of phosphate. In the presence of a *phoR⁺* allele the residual expression is still sensitive to phosphate expression (the expression is reduced to one percent), but in the presence of the *phoR⁻* allele the five percent expression of APase cannot be reduced by phosphate.

26. (a) Three kinds of repressor mutations might occur: (1) a repressor that cannot bind X, in which case the operon is on; (2) a repressor that cannot bind to operator even when X is present, in which case the operon is on; (3) a repressor that binds to the operator without X, in which case the operon is

off. (b) The phenotypes of partial diploids with wildtype and each of the mutants are: first mutant, inducible; second mutant, inducible; third mutant, unable to synthesize X under all conditions.

27. Post-translational modification is needed. Some possibilities are the following: (1) The active enzyme might be a multisubunit protein, and the monomer concentration takes time to reach a value at which the equilibrium shifts toward aggregation. (2) The protein must be modified by cleavage or addition of a sugar (i.e., it is a glycoprotein) before it is active. (3) A metal ion has to be bound.

CHAPTER 12

Genetic Engineering

CHAPTER SUMMARY

The recombinant DNA technique allows different DNA molecules to be joined and propagated indefinitely, thereby creating novel genetic units. Restriction enzymes play a key role in the technique, because they are able to cleave DNA molecules within particular base sequences. Fragments obtained from any two DNA molecules cleaved by a single restriction enzyme can be joined by annealing their complementary single-stranded ends. The carrier DNA molecule used to propagate a desired DNA fragment is called a vector; the most common vectors are plasmids, phages, and viruses. The $CaCl_2$ transformation procedure is an essential stage in propagation of recombinant molecules as it enables foreign DNA molecules to enter bacteria. A variation of the technique is applicable to eukaryotic cells. If the recombinant molecule has its own replication system or is able to use the system of the host, it can replicate. If the molecule is a plasmid, it can become permanently established in the host cell. If it is a phage, it can multiply and produce a stable population of phage carrying foreign DNA.

Several methods are used for joining DNA molecules--cohesive-end joining, blunt-end ligation, and homopolymer tail-joining. The first uses the complementary single-stranded ends produced by most restriction enzymes. The second utilizes the ability of phage T4 DNA ligase to join molecules without single-stranded ends. With the third a single strand of poly(dA) is added to one fragment and a single strand of poly(dT) is added to the other. Then, the single strands (the tails) are allowed to anneal. A cell whose properties have been changed by establishing a recombinant DNA molecule is said to have been genetically engineered. The recombinant DNA technique has been used to clone genes, primarily for the purpose of obtaining large quantities of either a gene (for use as a hybridization probe) or the gene product.

Several procedures have been used to obtained the original gene that is cloned. In some cases, the gene can be isolated from a restriction digest; occasionally, detection of the gene is straightforward, but sometimes ingenious and elegant methods, designed for the particular gene, are required. With eukaryotic genes a useful procedure is to isolate the mRNA derived from processing of the primary transcript. Such an RNA molecule can be converted to double-stranded DNA by the enzyme reverse transcriptase. Such DNA copies of an RNA molecule are called complementary DNA or c-DNA. They can be linked to various vectors either by homopolymer tail-joining or by adding to the termini of the c-DNA linkers that contain restriction sites. Treatment of the molecules with linkers with an appropriate restriction enzyme yields cohesive ends that can be used to join the c-DNA to a plasmid or phage that has been cleaved with the same restriction enzyme.

Synthesis of prokaryotic proteins in bacteria is

straightforward, but synthesis of eukaryotic proteins by
bacteria requires special procedures, such as linkage to
prokaryotic promoters and ribosome binding sites. Eukaryotic
proteins requiring posttranslational modification usually
cannot be synthesized in bacteria but some can be synthesized
in yeast. A few small eukaryotic proteins, for example, the
polypeptide hormones, can be obtained by cloning synthetic DNA
molecules, whose nucleotide sequence has been designed in
accordance with the genetic code to yield a desired amino acid
sequence.

Retroviruses and other viral vectors have been used to
transfer genes into animal cells; the possibilities of
modifying the properties of whole organisms, particularly
humans, in this way is called gene therapy.

A plasmid has been obtained from the bacterium
Agrobacterium tumefaciens, which infects many plants, and
this plasmid is being used to perform genetic engineering of
higher plants. Viral coat proteins synthesized in genetically
engineered bacteria will probably be a source of vaccines for
viral diseases. Harmless viruses have also been modified to
include thee coat proteins of harmful viruses, again as a
vaccine.

BOLD TERMS

alkaline phosphatase, blunt ends, c-DNA, cloned, cloning
vehicle, cohesive ends, colony hybridization assay,
complementary DNA, ethidium bromide, gene cloning, gene
libraries, gene therapy, genetic engineering, homopolymer
tail-joining, *in situ* hybridization assay, insertional
inactivation, linkers, lysozyme, Northern blotting,
palindromes, poly(dA) tails, probe, recombinant DNA
technology, restriction endonucleases, restriction enzymes,
restriction map, reverse transcriptase, Rous sarcoma virus,
Southern blotting, terminal deoxynucleotidyl transferase,
vector.

ADDITIONAL INFORMATION

The material on genetic engineering is straightforward and
should require little effort to understand it. Therefore, no
additional information is being provided. The student
interested in learning details about experimental techniques
should consult the Bibliography of *General Genetics.*

DRILL QUESTIONS

1. What properties of a cloning vehicle are essential?

2. Restriction enzymes generate three types of termini. What
   are they?

3. With the normal convention for orienting molecules, will a
   molecule with a 3' extension have the cut in the upper strand
   to the left or to the right of the cut in the lower strand?

4. Are the termini of a restriction fragment produced by a particular enzyme always the same or can they be different?

5. Can a restriction enzyme cut at two sites have different base sequences?

6. Can two different restriction enzymes act at the same site?

7. What features do all restriction sites have?

8. Will the sequences 5'GGCC and 3'GGCC be cut by the same restriction enzyme?

9. How is ethidium bromide used in genetic engineering?

10. Where do restriction enzymes come from?

11. What unique property is possessed by the enzyme reverse transcriptase, with respect to its polymerizing ability?

12. What is the advantage of using a plasmid with two antibiotic-resistance genes as a cloning vehicle?

13. What unique property is possessed by the enzyme terminal nucleotidyl transferase, with respect to its polymerizing ability?

14. What three methods can be used to join fragments with blunt ends?

15. How must a blunt-ended molecule be treated before homopolymer tail-joining can be carried out?

16. Of what use is *Agrobacterium tumefaciens* in genetic engineering?

17. Give two reasons why a cloned prokaryotic gene might not be expressed in a prokaryote.

18. What method may be used to avoid the problems in Question 17?

19. Give two reasons why a simply cloned eukaryotic gene will not usually yield functional mRNA in a bacterial host.

20. Assuming the problems in Question 19 are solved, give three reasons why a desired protein may not be produced from a eukaryotic gene cloned in a bacterium.

ANSWERS TO DRILL QUESTIONS

1. It must be able to replicate, it must be transformable, transformants must be selectable, and, preferably, it should have definite cleavage sites.

2. Blunt ends, cohesive ends with 3' extensions, and cohesive

ends with 5' extensions.

3. To the right.

4. Always the same.

5. No. There is a class of restriction enzyme that recognizes a particular site but makes cuts at random positions around the site; these enzymes are not used in genetic engineering.

6. Yes. See also Additional Problem 2.

7. They are palindromes.

8. No.

9. Supercoiled DNA molecules, such as plasmid DNA, have a different density from other DNA molecules when centrifuged to equilibrium in concentrated CsCl solutions containing ethidium bromide, so the supercoiled molecules can be easily purified.

10. They can be isolated from most, if not all, bacterial species.

11. It makes DNA from RNA.

12. One gene can be used to detect the plasmid in a transformation experiment, and if there is a restriction site in the other gene, lack of resistance to that antibiotic can be used to show that insertion has occurred.

13. It can add nucleotides to an extended single-stranded 3' terminus of a DNA molecule without the need of a template.

14. Blunt-end ligation, addition of linkers, homopolymer tail-joining.

15. Terminal nucleotidyl transferase will not add the tail to the blunt end. First, the DNA must be treated with a 5'-P-specific nuclease to remove a few nucleotides from the 5' end, leaving a 3' overhang.

16. *Agrobacterium tumefaciens* contains a plasmid that can be used to perform genetic engineering of plants. This should markedly improve the techniques of plant breeding.

17. It is always possible, but sometimes the gene may have been separated from its promoter or from its ribosome binding site.

18. A standard procedure is to design a vector so that it contains a strong promoter and a ribosome binding site (commonly that for the *E. coli lac* operon), with an adjacent downstream restriction site.

19. (1) The bacterium does not recognize a eukaryotic promoter. (2) If a primary transcript is actually made, it will

typically have introns and these cannot be removed in a bacterium.

20. (1) The mRNA lacks a ribosome binding site. (2) The protein must be processed. (3) The bacterium destroys the protein, having recognized it as foreign protein.

ADDITIONAL PROBLEMS

1. A DNA molecule is cleaved by a restriction enzyme and analyzed by gel electrophoresis. Only one sharp band is seen. Explain.

2. Restriction enzyme I cleaves in the sequence ATAT, between ATA and T and in the symmetric position in the complementary strand. Enzyme II cleaves all enzyme-I sites, but enzyme I cleaves only a few enzyme-II sites. Suggest an explanation.

3. The A^+ allele of a cell is easily selected by growth on a medium lacking substance A. However, repeated attempts to clone the A gene by digestion of cellular DNA and the vector by the enzyme EcoRI (the vector has an EcoRI sie) are unsuccessful. If HaeIII is used (the vector also has an HaeIII site), clones are easily found. Explain.

4. In cohesive-end joining why is DNA ligase needed to complete a joining reaction? In other words, since annealing has already joined the fragments and since replication can occur across a single-strand break, why must the joint be made a covalent one?

5. Can the c-DNA technique be used with any eukaryotic gene? This is a practical and not a theoretical question.

6. Suppose you had a plasmid with an *amp-r* gene having an EcoRI site and a *lacZ* gene with a SalI site. If you were trying to clone a particular gene, would it be easier to use EcoRI or SalI to cleave the DNA? Think about how you would detect the recombinant plasmid.

7. It is possible to eliminate a restriction site by a point mutation. However, is it possible, by mutation, to convert a restriction site for one enzyme to a restriction site for another enzyme?

ANSWERS TO ADDITIONAL PROBLEMS

1. There are three possibilities, of which the first two are rather unlikely: (1) A single cut is made precisely in the middle of the molecule. (2) Several cuts are made in positions such that all fragments have sizes that are indistinguishable by gel electrophoresis. (3) The molecule is a circle and a single cut is made.

2. The recognition site for enzyme I might be larger than ATAT, for example, GATATC, whereas the recognition site for

CHAPTER 12   177

enzyme II might be ATAT.

3. The *A* gene probably contains an EcoRI site, so the gene is destroyed. It does not contain an HaeIII site.

4. The cohesive ends are very short, so breathing will normally cause them to come apart at room temperature. In fact, the annealing reaction must sometimes be done below room temperature. The joint would not survive the higher temperatures (for example, 37 C) needed for transformation unless they were covalently joined. Without maintaining circularity, the DNA would be unable to complete its first round of replication within the host cell.

5. No, because a source of fairly pure mRNA is needed. This is really only practical with particular mRNA molecules made in very large quantities by particular cells, such as the cells making hemoglobin. If the mRNA is a minor species, there might be no way to isolate or identify it. Normally, one would do this by hybridization to a DNA molecule. However, often there is no way to obtain the DNA molecule except by cloning, and the c-DNA method may be the only way to do the cloning.

6. In principle, either could be used. If you used EcoRI, the plasmid could be detected by growth on medium using lactose as the sole carbon source. Colonies that grow could then be tested for antibiotic-sensitivity to find those contained DNA inserted in the *amp-r* gene. Note that this process would take two steps, and, in fact, two days of growth. If you used SalI, transformants with inserted DNA could be isolated in a single step by plating transformed bacteria on a lactose color-indicator medium containing ampicillin.

7. Not by a point mutation, because any change in a single base would destroy the palindromic symmetry of the sequence. One mutational change that can change a site in that way is a two-base deletion that removes the base pairs immediately on either side of the center of symmetry. A six-base site could be converted to a four-base site. However, such an event would be very rare and probably has not been observed.

SOLUTIONS TO PROBLEMS IN TEXT

1. (a) Same as a linear molecule, because free rotation at the site of the break would allow continued intercalation of the ethidium bromide molecules. (b) The density would be the average of the value for a covalent circle and that of the nicked circle.

2. The labeled fragments indicate the termini, which are 6.2 and 8.0 for EcoRI, and 6.0 and 10.1 for BamHI. Therefore, the BamHI map is 6.0, 12.9, 10.1. Let us now orient the EcoRI map with respect to the Bam HI map:

Figure 1

```
       6.0              12.9              10.1
   ─────────┼──────────────────┼──────────────────
```

If the 6.2-kb terminus of the EcoRI map were at the left, a 0.2-kb fragment would be in the double digest, but no such fragment is present. If the 8.0-kb terminus were at the left, a 2.0-kb fragment would be present, as it is. Thus, the 8.0-kb fragment is at the left, and the 6.2-kb fragment is at the right. Now, consider the 4.5-kb fragment. If it were next to the 6.2-kb fragment, the double digest would have an 0.6-kb fragment, which is not present. Also, if it were next to the 8.0-kb fragment, the double digest would have a 6.5-kb fragment, which it does not. Therefore, the 4.5-kb fragment cannot be next to either the 6.2-kb or the 8,0-kb fragment and must be in the center of the molecule. Now, consider the 7.4-kb fragment. If it were next to the 8.0-kb fragment, there would be a 2.5-kb fragment, which there is. If it were instead next to the 6.2-kb fragment, there would be a 3.5-kb fragment, which is not present. Thus, the 7.4-kb and 6.2-kb fragments are adjacent. Analysis of the position of the 2.9-kb fragment shows that it is next to the 8.0-kb fragment, which agrees with the position of the 7.4-kb fragment. Therefore, if the BamHI map has the order 6.0--12.9--10.1, the EcoRI map has the order 8.0--7.4--4.5--2.9--6.2, with the same orientation.

3. Note that the 3.6-kb and 5.3-kb fragments are now joined to form the 8.9-kb fragment. This suggests that they are terminal fragments of a linear molecule and that the intracellular DNA is circular.

4. (a) The plasmid is circular (that is, no free end), so all termini generated by the enzyme will have the same cohesive ends. (b) No. because the terminal fragments will have only one cohesive end and one natural end.

5. A mutation may alter one base in a restriction site and thereby cause two potential fragments to remain uncleaved.

6. (a) The *tet-r* gene has not been cleaved, so addition of tetracycline to the medium will require that the colonies be Tet-r and hence carry the plasmid. (b) Tet-r Kan-r and Tet-r Kan-s. (c) Tet-r Kan-s, because insertion will occur in the cleaved *kan* gene.

7. The fragments have eliminated either the promoters or the ribosome binding sites of both genes. The plasmid contains both of these, but they are on the same strand of the plasmid DNA. If genes A and B are transcribed in opposite directions, only one can be expressed from the plasmid.

8. A frameshift of two bases is generated, so all colonies will be Lac$^-$.

9. Select an enzyme whose restriction site is further from the *lac* promoter than the BamHI site. Let us assume it is HaeI. If the gene of interest does not contain a HaeI site do the following. Cleave the plasmid with both BamHI and HaeI and retain the fragment (there will be two) that contains the replication origin and other essential genes. One end of the fragment will have a BamHI terminus, and the other end will have a HaeI terminus. Cleave the donor DNA with both enzymes and join the fragments to the isolated plasmid fragment. Since each fragment has one BamHI terminus and one HaeI terminus, joining can only occur in a particular orientation.

10. The cloned fragment is flanked by tracts of G and of C, so the cleavage site of the enzyme must be the sequence GGG, GGGGG, or GGGGGG. Longer sequences of G are not expected, since restriction enzymes have not been observed with target sequences greater than about six nucleotide pairs.

11. A processed eukaryotic mRNA will generally possess a 3' poly(A) tail. Poly(dT) will anneal to this tail and prime DNA synthesis.

12. $2f = 2 \times 10^4/6 \times 10^9 = 3.3 \times 10^{-6}$. Therefore, $N = \ln(0.01)/\ln(1 - 3.3 \times 10^{-6}) = 1.38 \times 10^6$.

13. In order to become permanently established in *E. coli*, it must contain those elements necessary for autonomous replication in *E. coli*--namely, the replication origin and initiation proteins of the plasmid. For detection, it must carry antibiotic-resistance markers--preferably two markers with an insertion site in one of them, so that insertional inactivation can be used. In order to replicate at all in yeast, it must contain a replication origin. Recall in Chapter 4 that eukaryotes have multiple replication origins. In yeast, one of these origins is in the *trp* region, which accounts for the choice of that gene.

14. The principle is to design the system so that expression of the gene can be controlled. For example, a restriction enzyme could be used that separates the coding sequence and the promoter and the coding sequence could be linked to the promoter of the *lac* operon. Then, transcription could be turned on at will merely by adding an inducer of the *lac* operon.

15. Figure 2

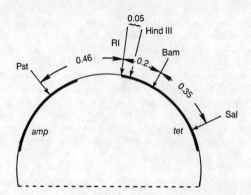

16. (a) Successful infection would be prevented by an alteration that would split a gene whose product is needed for replication and a rearrangement that would break the sequence in the replication origin. (b) Yes, as long as it does not introduce the changes described in part (a). (c) The answer is the same as in part (b).

17. A plasmid dimer.

CHAPTER 13

Somatic Cell Genetics and Immunogenetics

CHAPTER SUMMARY

Somatic cells are cells that are neither gametes nor precursors to gametes, and somatic cell genetics is the study of genetic processes that occur in such cells. In this chapter, hybrid somatic cells created by cell fusion and genetic events that occur during development of the immune system are examined.

A variety of techniques are available for fusing cells of like or different species. Hybrid cells can be selected easily if the two parental cell lines each carry a different mutation in the salvage pathway for DNA synthesis. The mutations complement in the hybrids and permit growth in HAT selective medium. Interspecies cell fusion is used in human genetics to assign genes to particular chromosomes. Fusion of a mouse cell with a human cell produces a hybrid cell with the full complements of 40 mouse chromosomes and of 46 human chromosomes. Human chromosomes in the newly formed hybrids are frequently lost; thus, from a population of hybrid cells, distinct lines of descendant cells, each line containing a different set of human chromosomes, usually results. When only a few human chromosomes remain in a cell, a hybrid becomes fairly stable and loss of human chromosome becomes infrequent. By studying these more-or-less stable hybrid cell lines a given human gene can be assigned to a particular chromosome, because its gene product will be made only in those hybrid cells that have retained the particular human chromosome containing that gene.

Individuals who have recovered from an infectious disease are less likely to become ill again with the same disease, because the initial exposure to the agent that causes the disease has made them immune to the organism. Immunity results from an acquired ability of the body to recognize and destroy invading viruses, bacteria, and large molecules. A substance able to elicit an immune response is called an antigen. The immune response relies on two types of white blood cells--B cells and T cells. B cells produce antibodies that circulate in the blood and lymph and combine with specific antigens; T cells attack specific antigens directly. Particular antigen-antibody reactions are not beneficial, resulting in blood incompatibility in transfusions and in hemolytic disease of the newborn. The antigens responsible for most types of transfusion incompatibility are the A and B antigens, whose presence on the surface of red blood cells is determined by codominant alleles. Adverse transfusion reactions usually result when antibodies in the blood of the recipient attack antigens on the transfused red blood cells. Hemolytic disease of the newborn results from the incompatibility of Rh blood groups between mother and fetus, when anti-Rh^+ antibodies in the mother can cross the placenta and destroy red blood cells of the fetus.

A normal mammal is able to produce more than one million different types of antibody molecules, but by using only a few hundred antibody genes. This variety results primary from DNA

splicing during development of B cells, which joins together
one each of numerous alternative coding sequences for three
parts of the light polypeptide chain of the antibody molecule.
The heavy polypeptide chain of the antibody molecule is
likewise formed by splicing of four segments. Additional
antibody variability originates from somatic mutations. Once
splicing and mutation has occurred, the cell containing the
new gene sequence multiplies, forming a clone of cells, each
of which is able to synthesize the same antibody. Each
antibody type made by an organism is the result of a unique
splicing event and is the product of a particular clone. At a
later time, if the organism is exposed to the same antigen,
the clone responds by making large quantities of the specific
anti-antigen antibody; this response is the cause of immunity.

Rejection of transplanted tissues or organs results from
attack by T cells of antigens present in cells of the
transplanted tissue. These antigens are determined by dominant
or codominant alleles, and an individual will possess all
antigens corresponding to the alleles present in the genotype
of the individual. Many genes produce tissue antigens that are
subject to rejection, but the gene resulting in the most rapid
and severe rejection reaction is the major histocompatibility
complex (MHC), which is a complex genetic region containing
several genes that participate in overall regulation of the
immune response. Many alleles of each of these genes occur,
and the particular combination of alleles that occurs in a
chromosome is called the haplotype of the chromosome. Some
diseases are associated with the MHC because particular
haplotypes are over-represented among affected individuals.

BOLD TERMS

aminopterin, antibodies, antigens, B cells, blood group
system, cell hybrid, cell lines, combinatorial joining,
compatible, constant regions, haplotype, HAT medium, heavy (H)
chain, histocompatibility antigens, hybridomas, immune, immune
response, incompatible, light (L) chain, J regions, major
histocompatibility complex, monoclonal antibody, rejection, Rh
negative, Rh positive, salvage pathway, somatic cell genetics,
T cells, V region, variable regions.

ADDITIONAL INFORMATION

Genetic experiments that use the mating techniques seen
throughout *General Genetics* cannot be done with humans,
for obvious reasons. For decades, genetic analysis relied on
the study of pedigrees in order to deduce the genotype of an
individual. Whereas this is still done in many cases, pedigree
analysis cannot always be done. One reason is that mutations
in genes that confer particular traits do not always exist.
Furthermore, some mutations produce changes that can only be
detected biochemically, and tissue from past generations is
rarely available. Cell fusion has provided the technique of
choice for mapping human genes without the need for mutations.
The mechanism of cell fusion is not well understood, but

understanding it is not necessary to use it as a tool. The feature of fusion that makes it a valuable technique is chromosome loss. That is, for unknown reasons, in the course of the first (roughly) 30 mitoses following fusion some chromosomes do not reach new daughter nuclei. In fusion between cells of different species, the chromosomes that are lost are predominately from just one of the species. (Actually, chromosomes might be lost from both species, but such a cell would be monosomic for a particular chromosome and would not survive because so many lethal mutations would be uncovered.) In the case of hybrids between humans and rodents it is the human chromosomes that are preferentially lost, a fact that is important for human genetics. The average number of human chromosomes remaining in a stable mouse-human hybrid is seven, with a range of 1 to 20. The significance of fusion is that biochemical activity and can be correlated with the presence of particular chromosomes. The method to locate a gene on a chromosome is simple: a set of hybrid cell lines, each having different human chromosomes, is assayed for a particular biochemical activity; the human chromosome present in all cell lines having the activity is then determined. For example, suppose cell lines A, B, and C contain the following human chromosomes: A, 1 and 2; B, 2 and 3; C, 3 and 4. If a gene were carried on chromosome 2, the gene product would be present only in lines A and B; if it were contained in chromosome 4, the gene product would be present only in cell line C. Conversely, if a gene product were present in cell line C, it could be on either chromosome 3 or 4. If the product were also present in cell line B, the gene would be on chromosome 3, and if it were not in cell line B, the location would be chromosome 4. This reasoning can also be used to discover the locations of several genes that contribute to the same phenotype or to production of the same protein, even when the genes are on different chromosomes. For example, consider the following cell lines: D, 1; E, 1, 2; F, 2. If a biochemical activity required the presence of a gene on chromosome 1 and another gene on chromosome 2, the activity would be present in cell line E, but not in either D or F. Conversely, if the activity were found in E, one could determine whether it was in chromosome 1 or 2 by examining cell line 1. If the activity were not in cell line A, it could be on chromosome 2. However, if the activity were also absent in cell line F, it could not be on chromosome 2 alone. A reasonable conclusion would be that at least two chromosomes are required for activity, and hence at least two genes. It should be noted that correlating an activity with a chromosome never indicates the number of genes required, for several necessary genes could be present on the same chromosome.

The nature of antigens is worth examining further, though it is not a genetic, but a biochemical topic. In *General Genetics*, an antigen is defined as a substance that elicits an immune response, and from the biology of the immune response it is clear that the antigen must be a foreign substance--that is, a substance that is not a normal component of the organism. We may ask, then, whether all foreign substances are antigens. For example, are all viruses antigenic, and is uranium oxide, certain a molecule not found in humans, also an antigen? Indeed, all viruses are antigenic, though different species do not have the same ability to elicit antibody production, and uranium oxide is not an antigen. It is probably fair to say that all foreign proteins are antigenic, as are most naturally occurring carbohydrates. Small molecules are never by themselves antigenic. However, under certain circumstances a small molecule can be helped to trigger an immune response, and thereafter, the organism will retain the ability to produce a specific antibody to the small molecule in all conditions. For example, the simple molecule dinitrophenol (a benzene ring to which is attached one OH group and two NO_2 groups) will not cause antibody production if injected in small quantities into a rabbit (in large quantities it is quite toxic). However, if dinitrophenol is

2,4-Dinitrophenol.

chemically linked to a simple protein, such as cattle albumin, which itself can cause anti-albumin to be made by a rabbit, and the dinitrophenol-albumin complex is injected into a rabbit, the rabbit will make an antibody that will react to dinitrophenol even when it is not linked to albumin. The immune system of the rabbit has been tricked into making an antibody that recognizes the complex, but the antibody will also recognize albumin alone and dinitrophenol alone. This technique does not work with all small molecules, but when it does, the small molecule is said to be a hapten. The general rule is that a hapten-protein complex will result in production of an antibody that can react both with the pure protein and with the pure hapten.

The student reading about the remarkable combinatorial joining that occurs during conversion of a germline cell to a B cell must surely ask him- or herself how it happens. There are two aspects to the question: (1) what is the biochemical mechanism for the splicing and (2) how is the germline cell instructed to produce the splicing pattern that will yield an antibody that will react with a particular antigen? Regretfully, these questions have not yet been answered; perhaps, in the next edition of *General Genetics* the answer can be given. It is worthwhile, however, to summarize the different features of the immune system that enables an organism to synthesize nearly 10^8 different antibodies: These features are the following: (1) Germline diversity: each haploid genome contains nearly 500 V, J, and D segments corresponding to the H and L chains; (2) combinatorial joining: nearly 800 different V-J units for L chains and 8000 different V-J-D units can be formed by splicing; (3) junctional variation: shifts in the junction points of the V-J and V-D-J junctions produce additional diversity; (4) somatic mutation: in a completely unknown way mutations occur in the V regions and these increase the possible number of antibody molecules; (5) random association of H and L chains: any pair of light chains can join with any pair of H chains to produce the two antigen-binding sites, which are jointly formed at the ends of the Y-shaped molecule by the V_L and V_H domains.

DRILL QUESTIONS

1. What is meant by a cell hybrid?

2. Does cell fusion ever occur spontaneously?

3. What methods are used to stimulate cell fusion?

4. What is probably the first step in cell fusion?

5. Once a cell hybrid has formed a single nucleus, what usually happens during subsequent mitoses?

6. When chromosomes are lost in the course of many mitoses, are the same chromosomes lost from every cell?

7. After cell fusion, does chromosome loss occur in most cells or in only rare cells?

8. How does one select or identify a mouse-human hybrid cell line containing particular human chromosomes?

9. Suppose four mouse-human cell lines (A, B, C, and D) have been isolated. Each contains one human chromosome, either 1, 2, 3 or 4, respectively. Cell line B possesses a human enzymatic activity not present in the other three cell lines. What information does this give you?

10. What is the significance of the enzymes HGPRT and TK in cell fusion?

11. What types of cells cannot grow in HAT medium?

12. What is the response of a body to an antigen?

13. Why is it important to match blood types with respect to the ABO system when transfusing blood?

14. Is transplantation rejection caused by pre-existing antibodies in the recipient?

15. Many molecules are antigens, but an antigen in one organism is not necessarily an antigen in another organism. What is the defining feature that makes a molecule an antigen for a particular organism?

CHAPTER 13 187

16. In the ABO blood group system which phenotypes have more than one genotype?

17. In the ABO system, which type of blood can be given to an individual with any other blood type, and why is that the case?

18. What blood types can a type O individual receive and why?

19. Which combination of Rh groups will cause hemolytic disease of the newborn? Can it occur in a firstborn child? Why?

20. What blood types could a child born of a type O mother and a type B father have?

21. How many chains and how many different types of chains are present in an IgG molecule?

22. What is meant by the variable and constant region of the individual chains of an IgG molecule?

23. Distinguish germ line cell and B cell.

24. Do the V and J segments of an antibody gene join to form the variable or the constant region of an IgG molecule?

25. How many genes are linked to form the H chain of an IgG molecule?

26. Which combination of relatives would form the best donor-recipient pair in tissue transplantation?

27. Where on a cell are the histocompatibilty antigens located?

ANSWERS TO DRILL QUESTIONS

1. A cell hybrid is a cell formed by fusing dissimilar cells.

2. Yes, but in the laboratory the frequency is usually increased by various means.

3. Treatment with Sendai virus, exposure to polyethylene glycol.

4. Intermingling of membranes.

5. Chromosomes are lost during subsequent mitoses.

6. No. This is the origin of the various cell lines.

7. Probably in all cells since cells that maintain both genomes are quite rare.

8. Cell lines are not usually selected. Instead, one examines many clones microscopically and classifies them according to the number of chromosomes of each type that are seen.

9. The gene encoding the enzyme is located in human chromosome 2.

10. Cell hybrids made by fusing two cells, each lacking one of these enzymes, can be isolated by growth in HAT medium.

11. A cell mutant in either the TK or HGPRT genes.

12. Synthesis of an antibody.

13. Antibodies in the blood can cause clumping of the red cells if blood of the wrong type is administered.

14. No. The immune system of the recipient recognizes the transplanted material as foreign and makes antibodies that destroy it.

15. A molecule can be antigenic only if the molecule is not a component of the organism. In certain disease conditions, known as autoimmune diseases, the immune system of an individual actually makes antibodies to molecular components of the individual.

16. Type A can be either I^A/I^A or I^A/O.
    Type B can be either I^B/I^B or I^B/O.

17. Type O blood, because the red blood cells have no antigens that can interact with any antibody in the blood of the recipient.

18. Only type O, because the type O individual makes both A and B antibodies.

19. Rh^+ father and Rh^- mother. The problem will be with a firstborn child. The cause of the disease is the following. An Rh^+ child developing in the uterus causes the mother to make antibodies to the Rh D antigen. In a subsequent pregnancy, these antibodies damage the blood cells of the fetus.

20. It depends on the genotype of the father, which could be I^B/I^B or O/I^B. Since one would not know which it was, the children could be either type O or type B.

21. There are four chains, two H chains and two L chains.

22. All IgG molecules do not have the same amino acid sequence. The constant region has nearly the same amino acid sequence in all IgG molecules, whereas the amino acid sequence of the variable region differs in each type of IgG molecule.

23. The germ line cell contains the nucleotide sequence found in sperm and egg. A germ line cell does not make antibody. The B cells, each of which makes a specific antibody, are the progeny of a germ line cell in which excision and splicing of segments of the IgG gene components has occurred.

24. Variable region. A V segment and a J segment are joined to the C region of the gene. The C region encodes the constant region.

25. Four. These are the V, J, and D segments, which are joined to the constant region.

26. Monozygotic twins, since their genotypes are identical.

27. On the cell surface.

ADDITIONAL PROBLEMS

1. Complementation can often be observed simply by growing mixed cultures of mutant cells, because of spontaneous fusion. This technique is sometimes used to understand the genetics of human metabolic disorders. A certain disorder is expressed in culture in that mutant cells fail to synthesize a particular antigen. Cells from six different human patients are grown in all possible paired combinations The results are shown in Table 1, in which a + indicates that the substance is made (which means that fusion has occurred) and a - means that it is not made.

	A	B	C	D	E	F
A	-					
B	+	-				
C	-	+	-			
D	+	+	+	-		
E	+	-	+	+	-	
F	-	+	-	+	+	-

TABLE 1

What do these results tell you about the genetic basis of the phenomenon?

2. A series of mouse-human hybrid cell lines are examined and found to have the following human chromosomes: (1) 4, 7, 8, 10, 17, X; (2) 3, 6, 21, X; (3) 9, 13, 19, 21, X; (4) 1, 4, 12, 19, X; (5) 2, 3, 13, 15, X. Human enzyme P is found in cell lines 3 and 4 only, and enzyme Q is absent from all of these cell lines. What do these results tell you about the location of the genes for these enzymes?

3. In a Gilbert and Sullivan operetta several maidens go to a particularly inept hospital to have their children and the midwives mix up four children. Fortunately, the captain of the ship is able to perform blood tests. Since the fathers are all known, he obtains the following results: Children, O, A, B, AB; parents: AB and O, A and O, A and AB, O and O. Match each child its parents.

4. A unmarried woman with blood type B, N, Rh$^+$, and who is a nonsecretor has child with O, MN, Rh$^-$ blood, who is a nonsecretor. She names a man as the father, who is a secretor

and has blood types A, N, Rh⁻. When his blood type is determined, his lawyer, who is a good geneticist, breathes a sigh of relief. What feature of the man's genotype produced this reaction?

5. A female strain-CH inbred mouse is mated with an inbred strain-C57 male. The F_1 hybrid is called CH57. From which of the following mice will a CH57 mouse tolerate skin grafts: CH, C57, CH57, a wild mouse, a mouse of another inbred strain?

6. Arrange the following relatives in order of their suitability as kidney donors to a 22-year-old male: his father, his brother, his uncle, his cousin, his mother, a monozygotic twin brother, a dizygotic twin sister.

7. It may seem farfetched, but the cells that differentiate to form antibody-producing cells have a property similar to that possessed by lysogenic phages and essential to the biology of these phages. What is this property?

ANSWERS TO ADDITIONAL PROBLEMS

1. There are three complementation groups: A, C, F; B, E, D. Thus, at least three genes are required to produce the phenotype. Note that this problem is no different from any other problem in complementation.

2. The only chromosome common to cell lines 3 and 4 is chromosome 19, so this must be the location of the genes encoding P. You know little about the location of the genes for Q other than that they are not in any of the chromosomes in the list.

3. O child: O and O parents; A child: A and O parents; B child: AB and O parents; AB child: A and AB parents.

4. His type-N blood makes it impossible for him to have fathered a MN child with a type-M woman.

5. A wild mouse and another inbred mouse will have, by chance alone, a large number of genetic differences, making a successful transplant very unlikely. Since CH and C57 and both inbred, they are homozygous, so the CH57 progeny will be a heterozygote, carrying each allele present in both CH and C57. Therefore, transplantation will be successful with CH, C57, and CH57.

6. Sex and age have nothing to do with the order, which instead is based on the number of common haplotypes and the probability of two individuals having identical haplotypes. The monozygotic twin is clearly optimal, as the genotypes will be identical. The child shares only one haplotype with a parent but siblings (brothers, sisters, or dizygotic twins) have a probability of 1/4 of having the identical haplotypes. Thus, the order is sibling, parent, uncle, cousin.

7. Both are able to engage in site-specific genetic exchange, in the one case between different genes that recombine to form IgG and in the other between attachment sites. This ability is actually quite rare.

SOLUTIONS TO PROBLEMS IN TEXT

1. Hypoxanthine and thymidine are substrates of HGPRT and TK, respectively. Aminopterin blocks the major pathway for synthesis of DNA precursors.

2. (a,b) 1 and 2 exclude the location of the gene as 6q or 11p and are informative. 5 correlates a deletion with absence of the gene and is informative, in fact, indicating that the gene is on chromosome 2. 3 and 4 provide no information. (c) The most likely candidates are the tip of 2p or the tip of 2q, as these tips are broken off and lost in formation of the ring chromosome. The true location of the ACP1 gene is near the tip of 2p.

3. (a) Some clones lack the human X chromosome, which indicates that there is no selection to maintain the HGPRT gene on the human X. Consequently, the human cells must have been TK^+. (b) Since all clones must retain the TK^+-bearing chromosome arm, the chromosome arm must be 17q.

4. For each enzyme, match the horizontal patterns in the table of cell lines with vertical patterns for each chromosome in the table of chromosomes. Let us consider steroid sulfatase. Each + and − match in the X-chromosome column except for the entries p and q. In clone F when only Xp is present, steroid sulfatase is also present. In clone E, when only Xq is present, the enzyme is absent. Thus, steroid sulfatase is in Xp. The locations of the other enzymes are the following: phosphoglucomutase-3, 6q; esterase D, 13q; phosphofructokinase, 21; amylase, 1p; galactokinase, 17q.

5. (a) As in problem 4, match this horizontal pattern of + and − with a vertical column in the table of chromosomes. The match is the column for chromosome 9. In clone A only q is present and the restriction site is present, so the chromosome and arm that carry the marker is 9q. (b) Since the ABO blood group marker is linked to the restriction site, it must also be in 9q.

6. As in Problems 4 and 5, the horizontal pattern matches column 1 in the table of chromosomes. Note that the locus is present when 1p is present (clone E) and absent when 1q is present (clone G). Thus, the location of the Rh locus is 1p.

7. (a) An *hh* individual cannot produce the precursor for either the A or the B antigens and hence is type O. (b) First cousins frequently share rare alleles. If both are carriers of the *h* allele, homozygous *hh* offspring, which are type O, can arise.

8. (a) Since the mother is a nonsecretor (a homozygous recessive), the father must be a secretor. Since the child is B and the mother does not carry I^B, that allele must have come from the father. The identity of the other allele of the father cannot be established from the single offspring, so he could be either AB or B. (b) Since the mother is se.se and the child is a secretor, the child must be heterozygous, Se.se. The mother is I^A/I^A or I^O/I^A and hence can form either I^A or I^O gametes. The father is I^A/I^B or I^O/I^B and can make three types of gametes, I^A, I^B, and I^O. The child is type B and cannot have the I^A allele, so it must have received the I^O allele from the mother. Thus, the father must have contributed the I^B allele, and the genotype of the child is I^B/I^O Se se.

9. (a) If the mother carried a D allele, she would not make antibodies to it and could not have a child with hemolytic disease. Therefore, her genotype is dd. (b) No. The woman already produces anti-D antibody. (c) 1/2, because dd fetuses are never at risk.

10. It cannot; otherwise the antibodies would produce maternal-fetal incompatibility with respect to the ABO blood groups. (Note: A few females do produce anti-A and anti-B antibodies of the IgG class and are at risk of severe maternal-fetal incompatibility.)

11. The offspring have the genotypes $xg\ xg$ (female) and Xg (male). Since the female is homozygous recessive, both parents must carry xg. Since the male is Xg, one parent must carry Xg; it cannot be the male, which already must carry xg, because the gene is X-linked. Therefore, the genotypes of the parents are $Xg\ xg$ (female) and $xg-$ (male).

12. The mother is genotypically $Xg\ xg$. Nondisjunction at the second meiotic division produces an egg cell with two X chromosomes of genotype $xg\ xg$, which is fertilized by a normal sperm that carries the Y chromosome.

13. (a,b) No, because an AB woman, which makes I^A and I^B gametes cannot have a type O (I^O/I^O) child, and MN, because an M woman cannot have an N child.

14. Male 1 is excluded because the baby must have received a $C3$ allele from its father. Male 2 is excluded because the baby must have received an M allele from its father.

15. Couple 1 cannot have an N baby, and couple 2 cannot have an Ee M baby. Thus, baby 1 belongs to couple 1, and baby 2 belongs to couple 2.

16. An identical twin, because identical twins must have the same genotype at all loci, including blood group and histocompatibility loci.

17. If the child carries only antigens inherited from its mother, then skin grafts from the child to the mother will be accepted. Otherwise they will be rejected because of antigens that are inherited from the father and that the mother does not possess.

18. No. Combinatorial joining is a random process producing different antibodies having the same or very similar specificities. The IgM molecules may differ (1) according to the V-J and V-D-J regions that became joined in the antibody-producing B cells and also (2) because of somatic mutation.

19. Apply the rule that one histocompatibility allele yields one cellular antigen. (a) Father and daughter share one haplotype. The probability that the sperm carries the haplotype present in the daughter is 1/2, and the probability that the egg carries the haplotype present in the father is 1/2. The shared haplotype is the only set of MHC genes common to father and daughter, so a homozygote can arise only by union of egg and sperm with the identical haplotype; this occurs at a frequency of (1/2)(1/2) = 1/4. (b) For the calf and bull to have the identical haplotype, the calf must receive the haplotype shared by father and daughter from the daughter, which is transmitted with a probability of 1/2, and the unshared haplotype from the father, which is also transmitted with a probability of 1/2. Thus, when father and daughter mate, the probability of a child that matches the grandfather perfectly is (1/2)(1/2) = 1/4.

20. Diploid, 2; triploid, 3; trisomic, 3, if the gene is in the trisomic chromosome, and, otherwise, 2.

21. (1) Examine the results in the display of page 487 of *General Genetics* and note that the probability for acceptance in the case P_2 donor and B_1 recipient is 1/2 for a single gene. Since there are three unlinked genes that assort independently, the probability is $(1/2)^3$. Similar reasoning is used for the other cases: B_1 donor, B_2 recipient, $(1/2)^3$; B_1 donor, P_2 recipient, 0; (4) F_2 donor, F_2 recipient, $(10/16)^3$.

CHAPTER 14

Population Genetics

CHAPTER SUMMARY

Population genetics is the application of Mendel's laws and other principles of genetics to populations of organisms. The population unit is a group of organisms of the same species living within a geographical region of such size that most matings occur between members of the group; this is called a local population or deme. In most natural populations, many genes are polymorphic in that they have two or more common alleles; one of the goals of population genetics is to determine the relative proportions of particular alleles (allele frequencies) and genotypes (genotype frequencies).

The relation between allele frequency and genotype frequency of a gene is determined in part by the frequencies with which particular genotypes form mating pairs; this is called the mating system for the gene. Three principal types of mating systems are (1) random mating, in which mating occurs independently of genotype, (2) assortative mating, in which mating pairs are more similar in phenotype (positive assortative mating) or less similar in phenotype (negative assortative mating) than would occur with random mating, and (3) inbreeding, in which mating pairs are genetically related.

When a population undergoes random mating for an autosomal gene having just two alleles, the frequencies of the genotypes are given by the Hardy-Weinberg rule. If the alleles of the gene are A and a, and their allele frequencies are p and q, respectively, then the Hardy-Weinberg rule states that the genotype frequencies with random mating will be: AA, p^2; Aa, $2pq$; and aa, q^2. The Hardy-Weinberg rule requires that there be no migration, differential survival or reproduction, or changes in allele frequency owing to chance. The Hardy-Weinberg genotype frequencies provide a convenient approximation to genotype frequencies that actually occur in many natural populations, and goodness of fit with Hardy-Weinberg frequencies can be evaluated with a chi-square test. The Hardy-Weinberg rule is easily extended to autosomal genes with multiple alleles or to X-linked genes. An important implication of the Hardy-Weinberg rule is that rare alleles occur much more frequently in heterozygotes than in homozygotes.

A local population undergoing random mating for a gene with two alleles is expected to have a frequency of heterozygotes of $2pq$. Large populations are often subdivided into smaller local populations with migration (but not random mating) occurring between the local populations. One effect of population subdivision is that the average proportion of heterozygotes among the local populations will be smaller than $2pq$, and this reduction is used in defining the fixation index as a quantitative measure of genetic divergence among the subpopulations.

Inbreeding means mating between relatives; the extent of inbreeding is measured by the inbreeding coefficient, the probability that the two homologous loci in an inbred individual carry alleles that are identical by descent. The

main consequence of inbreeding is that replicas of a rare
harmful allele present in a common ancestor may be transmitted
to both parents of an inbred individual and become
heterozygous in the inbred offspring. Among inbred
individuals, the frequency of heterozygous genotypes is
smaller, and that of homozygous genotypes greater, than would
occur with random mating.

Harmful recessive alleles are widespread in
Drosophila and other nonself-fertilizing populations. In
humans, the most important effect of inbreeding is that, when
compared with individuals resulting from random mating, inbred
individuals have a substantially increased risk of being
homozygous for rare harmful recessive alleles.

BOLD TERMS

allele frequency, allozymes, assortative mating, common
ancestors, consanguineous mating, Curly-Plum technique, deme,
F, fixation index, fixed, G, gene pool, genetic
differentiation, genetic divergence, genotype frequency,
Hardy-Weinberg rule, heterozygosity, identical by descent,
inbreeding, inbreeding coefficient, local population, mating
system, monomorphic, natural selection, path, polymorphic,
population, population genetics, population structure,
population subdivision, random genetic drift, random mating,
regular system, relative risk, subpopulations, total
heterozygosity, viability.

ADDITIONAL INFORMATION

All genetics is the genetics of populations. That is, one
cannot draw many interesting conclusions from experiments with
one flower or from the genotype of a single offspring.
However, in Mendel's experiments the populations were created
for the experiment, and experiments were usually set up with
organisms having clearly defined allelic frequencies. Natural
populations are subject to selection, so the frequencies of
dominant and recessive alleles at a single locus are almost
never identical. Population genetics is a discipline
established to analyze the distribution of alleles and
genotypes in natural populations and to make predictions about
how genotypic and phenotypic frequencies change from
generation to the next. It must be realized that for those
traits determined by a single locus, the rules of Mendelian
heredity are obeyed. Probably the most essential thing that a
student must learn is the completely elementary task of
determining an allele frequency. This will be reviewed here.

Consider a population of 100 individuals whose number and
genotypes are 37 *AA*, 52 *Aa*, and 11 *aa*. To
determine the allele frequency one must first count the
alleles. The 37 *AA* individuals contribute 74 *A*
alleles. The 52 *Aa* individuals contribute 52 *A* and 52
a alleles, and the 11 *aa* individuals contribute 22
a alleles. Addition yields the total number of alleles of
each type: 74 + 52 = 126 *A* alleles, and 52 + 22 = 74 *a*
alleles. Note that 126 + 74 = 200, which is twice the number

of individuals in the population. The allele frequency is just the fraction of the total number of alleles that consists of alleles of one type. Therefore, the allele frequency of A is just 126/200 = 0.63, and the allele frequency of a is 74/200 = 0.37. To ascertain that you have not made arithmetical errors, always check that the frequencies sum to 1; in this case, 0.63 + 0.37 = 1.00, so no mistake has been made.

The Hardy-Weinberg rule tells how genotype frequencies are distributed if mating is random; that is, p^2 AA, q^2 aa, and $2pq$ Aa, if p and q are the allele frequencies of the two alleles. If that is the case, mating is random. We can see whether the population described in the preceding paragraph is a result of random mating by calculating the genotypic frequencies using the Hardy-Weinberg rule. Note that we have just calculated that $p = 0.63$ and $q = 0.37$. Therefore, if mating is random, the frequency of homozygous dominants should be $p^2 = (0.63)^2 = 0.40$ and that of the homozygous recessive should be $(0.37)^2 = 0.14$. The frequency of heterozygotes is $2pq = 2(0.63)(0.37) = 0.46$. Again check your arithmetic by noting whether the three genotypic frequencies sum to 1. In this case, 0.40 + 0.14 + 0.46 = 1.00. Sometimes, because of differences in rounding off, the sum will be 0.99 or 1.01, but this should not cause concern. We now use these calculated frequencies to predict the number of individuals of each genotype if the population satisfied the Hardy-Weinberg rule. Since there were 100 individuals, the numbers are 40 AA, 14 aa, and 46 Aa. The observed values were 37, 11, and 52, respectively, which are not the expected values. Can we conclude that the population does not follow the Hardy-Weinberg rule? We cannot arbitrarily draw such a conclusion, but the chi-square test will provide information. The purpose of this test is to compare observed and expected data and use of the table in Chapter 1 of the text will yield the answer. The chi-square test must frequently be done in actual studies since, owing to statistical variation, observed and expected values are rarely identical.

DRILL QUESTIONS

1. How many A and a alleles are present in a population consisting of 10 AA, 15 Aa, and 4 aa individuals?

2. What is the relation between the number of alleles of a particular monogenic trait and the number of individuals in the population?

3. What are the allele frequencies in Question 1?

4. What relation must always hold for the frequencies of alleles at a particular locus?

5. What is meant by the statement that an allele is fixed in a population?

6. A population consists of 10,000 individuals. You want to know the allele frequencies but clearly will not screen every individual. Which number of individuals would you screen if you want a reliable value for the allele frequencies: 10, 400, 5000?

7. In what way do electrophoretic studies aid in detecting genotypic variation?

8. Is a dominant-recessive relation observed when using electrophoretic variation as a means of detecting variation?

9. The dominant alleles at three loci A, B, and C have allele frequencies of 0.62, 0.96, 0.85. Which of these loci is polymorphic?

10. At a particular locus in an animal the following genotypes are observed among 150 individuals: 58 AA, 79 Aa, 13 aa. What is the heterozygosity at the locus?

11. Six loci are studied in a plant. The heterozygosities are found to be 0.234, 0.652, 0.481, 0.129, 0.398, and 0.784. What is the value of the average heterozygosity?

12. Which of the following are probably examples of random-mating systems: humans, sea urchins (which are nonmotile animals that release sperm into ocean water), self-pollinating plants?

13. What are the genotypic ratios in random mating, if p and q are the allele frequencies for each of the alleles?

14. If the genotypic frequency of a homozygous dominant is 0.09, what is the frequency of the individual allele?

15. How does the frequency of a heterozygote compare to that of a homozygote for a rare allele?

16. For rare alleles, why are X-linked recessives expressed much more in males than in females?

17. How does the average heterozygosity differ from the total heterozygosity?

18. If a population is divided into two subpopulations and each has exactly the same allele frequencies, what will the fixation index G of the subpopulations be?

19. How does the frequency of heterozygotes in an inbred population compare to that in a random-mating population?

20. What is the inbreeding coefficient of a random-mating population?

21. Why is inbreeding generally to be avoided?

ANSWERS TO DRILL QUESTIONS

1. The AA contribute 20 A alleles, the Aa contribute 15 A and 15 a, and the aa contribute 8 a. Thus, there are 35 A alleles and 23 a alleles.

2. Each gene has two alleles, which may or may not be identical, so the number of alleles is twice the number of individuals.

3. There are 58 alleles altogether; 35, or 35/58 = 0.603 are A and 23 or 23/258 = 0.397 are a.

4. They must sum to 1.

5. The frequency of the allele is 1; that is, no other alleles for the particular gene are present in the population.

6. 10 is too small, for the statistical error will be too great; 400 is acceptable; 5000 is certainly unnecessary, though it will certainly be more accurate than the values obtained from 400 individuals.

7. The biological activity of an altered (mutant) protein is not always detectably different or even affected by an amino acid change. However, the electrophoretic mobility is very often affected by a change in one amino acid. Thus, electrophoretic differences are phenotypic differences and reflect genotypic differences. Many genotypic variants will of course be missed though.

8. No, because a protein is made from both copies of the gene (the alleles are codominant).

9. By definition, loci A and C are polymorphic, because the most common allele occurs at a frequency of less than 0.95.

10. 79/150 = 0.527.

11. The sum of the numbers is 2.678. Dividing by 6 yields 0.446.

12. Only the sea urchin, because males release their sperm into the water, where they drift randomly until encountering eggs. The male sea urchin does not even know who he is fertilizing.

13. For the homozygous individuals, p^2 and q^2, and $2pq$ for the heterozygous individuals.

14. The square root of 0.09, or 0.30.

15. The heterozygote is much more common than the homozygous recessive.

16. The female can be heterozygous and in fact the allele in a female will almost always be in the heterozygous state and hence unexpressed. The male cannot be heterozygous for an X-linked allele since the allele is always expressed.

17. The average heterozygosity of several subpopulations is the average value of the frequencies of heterozygotes in each of the populations. The total heterozygosity is the frequency of heterozygotes in a hypothetical population whose allele frequencies are the averages of the allele frequencies of the individual populations.

18. Zero.

19. It is smaller by the amount $2pqF$, in which F is the inbreeding coefficient.

20. Zero.

21. Inbreeding allows a higher frequency of production of individuals homozygous for rare deleterious alleles, and these individuals often have major defects.

ADDITIONAL PROBLEMS

1. If the genotype frequency for the homozygous dominant is 0.073, what are the frequencies of the homozygous recessive and the heterozygote, if the Hardy-Weinberg rule is obeyed?

2. Coat color of the pterodactyl is determined by a pair of sex-linked codominant alleles B and W. BB or $B-$ animals are brown, WW or $W-$ are white, and BW are patched. A population of pterodactyls had the following phenotypes:

   Males: 537 brown, 199 white
   Females: 489 brown, 31 white, 172 patched.

   (a) What are the allele frequencies?
   (b) What are the phenotypic frequencies of the next generation for each sex?

3. In an isolated population of 1600 people all individuals have type O blood. In another isolated population all people have type A blood. If 400 people from the second population move into the first, what will be the frequencies of the blood groups after one generation of random mating?

4. What are the frequencies of genes A and a if 36 percent of the population is heterozygous?

5. Carriers of Tay-Sachs disease were found to occur at a frequency of 0.07 in one city. What frequency of matings would produce an affected child, and what fraction of the children will develop Tay-Sachs disease in adulthood? Assume a penetrance of 1.

6. In the United States the frequency of the I^O blood group allele is 0.67 and that of the D allele (determining the Rh phenotype) is 0.6. What fraction of the population is both type O and Rh$^+$?

7. A particular disease is a result of a recessive gene, which in the homozygous state has 100 percent penetrance. In a particular country 1 person in 3500 has the disease. Assuming that people with the disease are always able to breed, how many carriers are there, if the population of the country is 50 million?

8. In a population of randomly mating animals 1 animal in 1000 is a homozygous recessive. A disease kills all of the homozygous recessives in one generation. What fraction of the population will be homozygous recessives after one generation of random mating?

9. Which alleles in humans would be more likely to follow the Hardy-Weinberg equation: the blood-group alleles or hair color?

ANSWERS TO ADDITIONAL PROBLEMS

1. The allele frequency for the dominant is $(0.073)^{1/2} = 0.27$. Since allele frequencies must sum to one, the frequency of the recessive allele must be 0.73, and the genotype frequency for the homozygous recessive is $(0.73)^2 = 0.533$. The frequency of the heterozygote is $1 - (0.073 + 0.533) = 0.394$.

2. (a) The total number of B alleles is $537 + 2(489) + 172 = 1687$, and the total number of alleles (X chromosomes) is $537 + 199 + 2(489 + 31 + 172) = 2120$. Thus, the frequency of the B allele is $1532/2120 = 0.796$, and the frequency of the W allele is $1 - 0.796 = 0.204$. (b) The allele frequencies of the F_1 males will equal the allele frequencies of the parental females: $[2(489) + 172]/1384 = 0.831 = p$ and $[2(31) + 172]/1384 = 0.169 = q$. The frequency of the F_1 females is determined from the following table:

		Frequency of parental males	
		0.73 B	0.27 W
Frequency of parental females	0.831 B	0.607	0.224
	0.169 W	0.123	0.046

3. There are two possible genotypes for type A blood, but since the second population is entirely type A, all of the people must be homozygotes; otherwise, some would be type O. Therefore, in the newly mixed population, the allele frequencies are $I^A = 0.2$ and $I^O = 0.8$. After one

generation of random mating the genotype frequencies would be: $I^A I^A$, $(0.2)(0.2) = 0.04$; $I^O I^O = (0.8)(0.8) = 0.64$; $I^A I^O = 2(0.2)(0.8) = 0.32$. The blood groups are: O, 0.64; A $(0.04 + 0.32) = 0.36$.

4. $2pq = 0.36$, so $pq = 0.18$. It is always true that $p + q = 1$, or $q = 1 - p$, so $p(1 - p) = 0.18$. This is a quadratic equation whose solutions are $p = 0.765$ or $p = 0.235$. Thus, the gene frequencies are either 0.765 A and 0.235 a, or 0.235 A and 0.765 a. Note that it will always be the case that there are two sets of frequencies for each value of the heterozygous fraction.

5. An affected child can only result when two heterozygotes (carriers) mate, and this will occur in only $(0.07)(0.07) = 0.0049$ of the matings. The allele frequency is half the carrier frequency, or 0.035. Thus, if random mating occurs, the frequency of children who will develop the disease is $(0.035)(0.035) = 0.00125$, or 1 in 800.

6. A type O individual is homozygous, so the frequency is $(0.67)^2 = 0.45$. An Rh^+ individual can be either homozygous or heterozygous. If the frequency of D is 0.6, that of d is 0.4, and the phenotypic frequency of Rh^+ is $(0.60)^2 + 2(0.6)(0.4) = 0.84$. The frequency of O Rh^+ is the product of these frequencies, or $(0.45)(0.84) = 0.38$.

7. Since the affected people are homozygous recessive, the frequency of the allele is $(1/3500)^{1/2} = 0.017$. The frequency of the dominant allele is 0.983. The frequency of carriers is $2(0.017)(0.983) = 0.033$, and the number of carriers is $(0.033)(50\ \text{million}) = 1.67$ million.

8. Before disease struck, the allele frequency for the recessive allele was $(0.001)^{1/2} = 0.032$, so the frequency of the dominant allele was 0.968. The genotype frequencies were 0.937 homozygous dominant and $2(0.032)(0.968) = 0.062$ heterozygous. After the killing the frequency of heterozygotes was $(0.062/(0.062 + 0.937) = 0.062$. Homozygous recessives are produced only when heterozygotes mate, and 1/4 of the progeny are recessives. The fraction of progeny that are homozygous recessives is $(0.062)(0.062)/4 = 0.00096$. Note that the value is not very different from the original frequency. This is because the frequency of homozygous recessives was initially quite low, so matings between homozygous recessives did not contribute much to the production of homozygous recessive progeny.

9. Blood groups. Hair color is often a criterion for selecting a mate, so random mating would probably not exist for hair color.

SOLUTIONS TO PROBLEMS IN TEXT

1. In an FF homozygote only F monomers are made, so only a single form, the F + F dimer will be present. The same

reasoning applies to an SS homozygote, which produces only an S + S dimer. For an *FS* heterozygote, both F and S monomers are present, and these will dimerize independently to form three types of dimers, an F + F dimer, a S + S dimer, and an F + S dimer.

2. The number of alleles of each type are: *a*, 2 + 2 + 5 + 1 = 10; *b*, 5 + 12 + 12 + 10 = 39; for *c*, 10 + 5 + 5 + 1 = 21. The total number is 70, so the allele frequencies are: *a*, 10/70 = 0.14; *b*, 39/70 = 0.56; and *c*, 21/70 = 0.30.

3. (a) The number of *a* alleles is 17 and that of *b* alleles is 53. Thus, the allele frequencies are: *a*, 17/70 = 0.24; *b*, 53/70 = 0.76. (b) The genotype frequencies are: *aa*, $(0.24)^2$ = 0.057; *bb*, $(0.76)^2$ = 0.578; *ab*, 2(0.24)(0.76) = 0.365.

4. The frequency of the recessive allele is $q = (1/14,219)^{1/2}$ = 0.0084. Thus, p = 1 − 0.0084 = 0.9916, and the frequency of heterozygotes is $2pq$ = 0.017 (about 1 in 60).

5. (a) If 0.60 can germinate, 0.40 cannot, and these must be homozygous recessives. Thus, the allele frequency q of the recessive is $(0.4)^{1/2}$ = 0.63. Then, p = 1 − 0.63 = 0.37 is the frequency of the resistance allele. (b) The frequency of heterozygotes is $2pq$ = 2(0.37)(0.63) = 0.46. Since the 0.60 that could germinate are either heterozygous or homozygous dominant, the frequency of homozygous dominants is 0.60 − 0.46 = 0.14, and the proportion of the surviving genotypes that is homozygous will be 0.14/0.60 = 0.23.

6. If 0.85 are Rh⁺, then 0.15 are Rh⁻, and the probability of a mating between an Rh⁺ individual and an Rh⁻ individuals is (0.85)(0.15) = 0.13.

7. The frequency of homozygous recessives is 0.10, so the frequency of the recessive allele $q = (0.10)^{1/2}$ = 0.316. The frequency of the dominant allele is 1 − 0.316 = 0.684, so the frequency of heterozygotes is 2(0.316)(0.684) = 0.43. also, 0.90 of the individuals are nondwarfs, so the frequency of heterozygotes among nondwarfs = 0.43/0.90 = 0.48.

8. $(1/2)p + (1/2)q$. However, since $p + q = 1$, this equals 1/2, independently of the allele frequency.

9. (a) The allele frequencies or *A* and *a* are 0.8 and 0.2 in males and 0.2 and 0.8 in females. Remember that only males and females mate. Thus, after one generation of random mating the genotype frequencies in both sexes are (0.8)(0.2) = 0.16 *AA*, (0.8)(0.8) + (0.2)(0.2) = 0.68 *Aa*, and (0.2)(0.8) = 0.16 *aa*. The allele frequency of *A* is 0.16 + (1/2)(0.68) = 0.50; that of *a* is the same, and the frequencies are the same in both sexes. However, the genotypes are not in Hardy-Weinberg proportions. (b) One additional

CHAPTER 14   205

generation of random mating results in genotype frequencies of (0.5)(0.5) = 0.25 AA, 2(0.5)(05) = 0.50 Aa, and (0.5)(0.5) = 0.25 aa, which are in Hardy-Weinberg proportions.

10. (a) No. A homozygote for self-sterility alleles cannot occur, because it could only be formed by a sterile mating (A x A or a x a), which is not possible. (b) If there were two alleles, a and b at the locus, the style would have the genotype ab (homozygotes cannot be formed, as just explained), and the pollen could either carry the a or b allele, neither of which could germinate. If there was a third allele, c, an ac plant could make a c pollen grain that could germinate of the style of an ab plant. Thus, three alleles is the minimum number required for fertility.

11. The baldness (b) allele has a frequency of 0.3, so the frequency of the nonbaldness (n) allele is 0.7. In females, the only genotype that has a bald phenotype is bb, which occurs at a frequency of (0.3)(0.3) = 0.09. Nonbald females occur at a frequency 1 - 0.09 = 0.91. In males the only nonbald genotype is nn, whose frequency is (0.7)(0.7) = 0.49, so the frequency of bald males is 1 - 0.49 = 0.51.

12. The frequency of the recessive is 0.2. Thus, for females the genotypic frequency of the homozygous recessive (yellow body) is $(0.20)^2$ = 0.04 and the number of such flies is (0.04)(1000) = 40; thus, 1000 - 40 = 960 is the number of wildtype flies. Because the gene is X-linked, the genotypic frequency for the recessive in males is the allele frequency, or (0.20), and the number of yellow males is (0.20)(1000) = 200. The number of wildtype flies is 1000 - 200 = 800.

13. I^A/I^A, (0.16)(0.16) = 0.026; I^A/O, 2(0.16)(0.74) = 0.236 I^B/I^B, (0.10)(0.10) = 0.01; I^B/O, 2(0.10)(0.74) = 0.148; O/O, (0.74)(0.74) = 0.548; I^A/I^B, 2(0.16)(0.10) = 0.032. The phenotype frequencies are: O, 0.548; A, (0.026 + 0.236) = 0.262; B, (0.01 + 0.148) = 0.158; AB, 0.032.

14. The number of M and S alleles are 195 and 73 respectively, so the allele frequencies are 0.728 M and 0.272 S, respectively. The expected genotype frequencies are: (0.728)(0.728) = 0.530 MM, 2(0.728)(272) = 0.396 MS, and (0.272)(0.272) = 0.074 SS. Multiplying by 134 yields the numbers: 71.0 MM, 53.1 MS, and 9.9 SS, respectively. Calculation of chi-square gives a value of 1.8, which is nonsignificant; therefore, the observed numbers are not inconsistent with random mating.

15. The number of L and M alleles are 401 and 279 respectively, to which correspond the allele frequencies 401/680 = 0.590 and 279/680 = 0.410. The expected genotype frequencies are (0.590)(0.590) = 0.348 LL, 2(0.590)(0.410)

= 0.484 LM, and $(0.410)(0.410) = 0.168$ MM. Multiplying by 340 gives the expected numbers: 118.3, 164.6, and 57.1, respectively. Comparing observed and expected numbers gives a value of chi-square of 11.0, which is highly significant. Therefore, the observed genotype frequencies are not in Hardy-Weinberg proportions, because these marsh frogs do not constitute a single large random mating population but are split up into smaller local breeding populations.

16. (a) Use Equation 14-6 for the homozygous recessive. The allele frequency is the square root of the frequency among unrelated offspring $(8.5 \times 10^{-6}) = 2.9 \times 10^{-3}$. For $F = 1/16$, the frequency of homozygous recessives is $(8.5 \times 10^{-6})(1 - 1/16) + (2.9 \times 10^{-3})(1/16) = 1.90 \times 10^{-4}$; for $F = 1/64$, the value is 5.40×10^{-5}. (b) For first cousins, by an increase of a factor of $(1.9 \times 10^{-4})/(8.5 \times 10^{-6}) = 22.3$; for second cousins, by an increase in a factor of $(5.40 \times 10^{-5})/(8.5 \times 10^{-6}) = 6.4$.

17. $(0.01)(1/16) + (0.99)(0) = 0.0006$.

18. Use Equation 14-7. For individual I, $n = 5$. Thus, for $F_A = 0$ the inbreeding coefficient is $(1/2)^5 = 0.031$; for $F_A = 1/16$ the inbreeding coefficient is $(1/2)^5(1 + 1/16) = 0.033$.

19. Use Equation 14-7. $(1/2)^3 + (1/2)^3 + (1/2)^5 + (1/2)^5 + (1/2)^5 + (1/2)^5 = 3/8$.

20. (a) Use Equation 14-6 with $F = 0.66$, $p = 0.43$, and $q = 0.57$. The genotype frequency for aa is $(0.43)^2(1 - 0.66) + (0.43)(0.66) = 0.347$; for ab, $2(0.43)(0.57)(1 - 0.66) = 0.166$; for bb, $(0.57)^2(1 - 0.66) + (0.57)(0.66) = 0.486$. (b) With random mating the frequencies are: aa, $(0.43)(0.43) = 0.185$; ab, $2(0.43)(0.57) = 0.490$; bb, $(0.57)(0.57) = 0.325$.

21. Use Equation 14-5 with $H = 0.12$, $p = 0.8$, and $q = 0.2$. $F = 1 - 0.12/(2)(0.8)(0.2) = 0.625$.

22. (a) The heterozygosity in the population in southern Israel is $2(0.87)(0.13) = 0.226$, and that in northern Israel is $2(0.24)(0.76) = 0.365$. The average heterozygosity H_S is $(1/2)(0.226 + 0.365) = 0.295$. By fusing the populations the average allele frequencies would be $(1/2)(0.87 + 0.24) = 0.555$ and 0.445, so the total heterozygosity H_T would be $2(0.555)(0.445) = 0.494$. Thus, by Equation 14-4 $G = (0.494 - 0.295)/0.495 = 0.40$. (b) From the chart below Equation 14-4 and the graph in Figure 14-9, the divergence is very great. That is, the populations have undergone very great genetic divergence with respect to the alleles at this locus.

23. The heterozygosities were $2(0.59)(0.41) = 0.484$ and $2(0.97)(0.03) = 0.058$, so $H_S = (1/2)(0.484 + 0.058) =$

0.271. The average allele frequencies are $(1/2)(0.59 + 0.97) = 0.78$ and $(1/2)(0.41 + 0.03) = 0.22$, so $H_T = 2(0.78)(0.22) = 0.343$. By Equation 14-4, $G = (0.343 - 0.271)/0.343 = 0.21$, corresponding to great genetic divergence.

CHAPTER 15

Genetics and Evolution

CHAPTER SUMMARY

Evolution is the progressive increase in the degree to which a species becomes adapted to its environment. A principal mechanism of evolution is natural selection, in which individuals superior in survival or reproductive ability in the prevailing environment contribute a disproportionate share of genes to future generations, thereby gradually increasing the frequency of the favorable alleles in the whole population. However, at least three other processes can also change allele frequency--mutation (heritable change in a gene), migration (movement of individuals among local populations), and random genetic drift (resulting from restricted population size).

Spontaneous mutation rates are generally so low that the effect of mutation on changing allele frequency is minor. However, continuously growing bacterial cultures are sufficiently large that the effects of mutation can be detected and measured.

Migration can have significant effects on allele frequency because migration rates may be very large. The main effect of migration is the tendency to equalize allele frequencies among the populations that exchange migrants.

Selection occurs through differences in viability (the probability of survival of a genotype), and in fertility (the probability of successful reproduction). The viability and fertility of a genotype are usually measured and expressed relative to the viability and fertility of some other genotype in the same population. The net contribution of a genotype to the offspring of the next generation, taking both viability and fertility into account, constitutes the fitness of the genotype.

Populations maintain harmful alleles at low frequencies as a result of a balance between selection, which tends to eliminate the alleles, and mutation, which tends to increase their frequencies. The equilibrium allele frequency that occurs with selection-mutation balance is usually significantly greater for alleles that are completely recessive than for alleles that are partially dominant. This difference arises because selection is quite ineffective in affecting the frequency of a completely recessive allele when the allele is rare, owing to the almost exclusive occurrence of the allele in heterozygotes.

A few examples are known in which the heterozygous genotype has a greater fitness than either of the homozygous genotypes (favored heterozygote). This case is called overdominance, and it results in an equilibrium in which both alleles are maintained in the population. An example of overdominance is sickle-cell anemia in equatorial Africa where falciparum anemia is endemic. Individuals homozygous for the normal hemoglobin beta allele are not anemic but are susceptible to malarial infection, and individuals homozygous for the mutant beta gene are severely anemic and usually die very young. However, heterozygous individuals have an

increased resistance to malaria and only a mild anemia, which results in greater fitness.
    Random genetic drift is a statistical process of change in allele frequency in small populations, resulting from the inability of every individual to contribute equally to the offspring of successive generations. In a subdivided population random genetic drift is a principal cause of divergence in allele frequency. In an isolated population, barring mutation, an allele will ultimately become fixed or lost as a result of random genetic drift. However, the probability of ultimate fixation of an allele is equal to the frequency of the allele in the initial population.

BOLD TERMS

adaptation, additive, effective population number, equilibrium frequencies, fertility, forward mutation, heterozygote superiority, irreversible mutation, migration, migration rate, mutation, one-way migration, overdominance, random genetic drift, relative fitness, selection, selection coefficient, selectively neutral, viability.

ADDITIONAL INFORMATION

In the experiment on mutation from sensitivity to resistance to phage T5, described in *General Genetics,* the mutation was considered to be irreversible. It was pointed out that considering it so is a matter of mathematical approximation, because, in fact, the reverse mutation does occur. It is worth noting that there are mutagenic events that are truly irreversible, in the sense that the probability of reversibility is so low that there is no reason not to consider them irreversible. A deletion is such a mutation, because the probability of replacement of the precise sequence deleted is extraordinarily small. It was also stated that mutation is a weak force in evolution, even though it is the ultimate source of variation. The meaning of this can be made clear by considering the change from sensitivity to resistance to T5 in nature, rather than in the laboratory. As pointed out in the text, after 1000 generations the frequency of T5-sensitive bacteria had decreased from 1.0 to 0.999922, and after one million generations it would be 0.92. This large number of generations represents about 65 years of continuous growth in the laboratory. However, in nature, bacteria do not grow continually, for a variety of reasons. A typical bacterium probably engages in only about 100 doublings per year, so 10^6 generations would take about 10^4 years. However, in much less a population of *E. coli* would probably encounter a T5 phage, in which case in a matter of hours only T5-r cells would remain. This rapid conversion of phenotype is of course a rather extreme example of selection.
    The term selection coefficient requires some care, because the word coefficient so often refers to a multiplicative factor. In this case, the selection coefficient for a particular genotype is defined as the difference between the fitness of that genotype and the fitness of a genotype

chosen as a standard. Let us use the concept to perform a simple practical calculation. Assume that you have, by chance, isolated a yeast mutant that grows 10 percent more rapidly than the wildtype and that you would like to have several of these mutants to study. However, you have no way to detect it other than by a tedious measurement of growth rate. Clearly it would be unreasonable to pick clones at random and measure the growth rate unless the number of mutants is about 1 in 10. In the absence of any selective procedure, a culture derived from a single cell would have perhaps one fast-growing mutant per 10^7 cells. By simply growing the culture, the fraction of cells that grow more rapidly should increase, and we can calculate how many generations are required to reach a frequency of 0.1. A mutant growing 10 percent more rapidly will have a fitness of 1.10 compared to the standard. We use the rearranged form of Equation 15-5 given in the second display on page 542, with $p_0/q_0 = 10^{-7}$ and $p_n/q_n = 0.1/0.9$. Thus, $n = (1/\log 1.10)[\log (0.1/0.9) - \log 10^{-7}] = 146$ generations. That is, after 146 generations about 1 yeast cell in 10 should have a generation time 10 percent higher than the original cells.

In calculating the effects of migration the critical values are the allele frequencies in the source and recipient populations. The difficulty in analyzing the effects of migration is knowing when migration started and what the allele frequency was at the time. In general, one will not know these numbers, though in an experimental situation this information is usually available. The simplest real case to analyze is that of introduction of a totally new allele into a population. In applying Equation 15-4 to the population it is necessary to recognize that p_0, the allele frequency of the recipient population before the onset of migration, is zero. Furthermore, if the donor population is homozygous, the allele frequency for that population is 1. This case is examined in one of the problems in the text.

Random genetic drift is the change in gene frequency from generation to generation owing to sampling variation. The magnitude of the change is inversely proportional to the size of the population. The reasoning is straightforward, though somewhat abstract, and is perhaps best understood by an example. Consider two populations, each consisting of four animals, a BB male, a Bb male, a Bb female and a bb female. The allele frequency of each allele is 0.5 in both populations. In population 1, two matings occur—BB x Bb, and Bb x bb. The first yields one BB child, and the second two bb and 1 Bb. The allele frequencies are then 0.375 B and 0.625 b. In the second population the matings BB x bb and Bb x Bb produce 1 BB and 2 Bb, respectively, yielding an allele frequency for B of 0.625. Note that the differences between the allele frequencies of the progeny are a result of chance—which pairs mated and how many progeny there were. In a second round of

mating in each population the allele frequencies might move nearer to 0.5 or further away, again due to chance. These variations in gene frequency from one generation to the next are what is referred to as random genetic drift. The critical point is that alleles can be lost. For example, in the first population there might have been only one mating $BB \times Bb$ yielding only two progeny, both BB. The allele frequency of B among the breeding population would then be 1, and we would say that the B allele is fixed and the b allele is lost. Henceforth, the allele frequencies in this population would be constant. If the populations had initially not consisted of just four individuals, but many millions, differences in gene frequencies in progeny of particular matings would compensate for one another. That is, there would be some populations in which there were more B and some with more b, and in the limit of an infinitely large population, the allele frequency of B would remain 0.5.
This is random genetic drift: the change in allele frequencies that result from chance sampling of gametes from parents. The effect is that some populations will ultimately be fixed for one allele or the other, because if a population ever arises for which an allele frequency is 1 or 0, the allele frequency of that population can never change, because the other allele is no longer present. Clearly, the larger the population, the more generations will be required for fixation or loss to occur. What this all means on a worldwide scale is that this type of statistical process is one reason for the existence of different allele frequencies among subpopulations. Two other factors, which are variants of genetic drift, also account for these differences. One is called the founder effect and refers to the fact that certain populations were initiated by a small number of individuals. For instance, some religious groups, such as the Amish in Pennsylvania, have developed from a small number of original settlers arriving from Germany. Random genetic drift in this population has had the effect that frequencies of a variety of alleles--for example, the MN blood group, differ significantly from that found in the region of Europe from which they came. Another effect is called the bottleneck effect. This refers to a catastrophic decrease in the size of a population (caused by famine, disease, natural disasters, wars, etc.) that cause random killing of the breeding population and hence change allele frequencies in one step. Random genetic drift acting on the resulting small population can cause significant changes in allele frequency from the original values. The founder and bottleneck effects are considered by some students of evolution to be the major factors responsible for differences in allele frequencies among natural populations in various parts of the world.

An important feature of random genetic drift is that all members of a particular generation are not always mating, and if they are, they are not participating in the formation of progeny on an equal basis. These facts are responsible for the notion of effective population size, as described in the text. This concept is designed to enable one to think of each set of progeny as arising in a standard way from a theoretically ideal population. Simply speaking, the effective population size for any generation is the number of parents that produce the next generation.

The student may wonder what the concepts in this chapter are for; that is, what are geneticists trying to understand when talking about selection, fitness, random genetic drift, and so forth? Basically, there are three goals: to understand the reason for differences in allele frequency in various populations; to predict how allele frequencies might change under the influence of mutation, migration, and environmental effects; and to understand the mechanisms by which different species evolved. The third point is one of the major questions in theoretical genetics and certainly the most important question in modern study of evolution.

DRILL QUESTIONS

1. What are the four major causes of changes in allele frequencies in natural populations?

2. Which of the four causes in Question 1 is affected by population size?

3. What is meant by a population being in equilibrium with respect to a particular mutation?

4. What is the expression for allele frequency for a particular allele if the population is in mutational equilibrium?

5. What are the units of m, the migration rate?

6. Can fitness be greater than 1? Can relative fitness be greater than 1?

7. In a population of diploid organism, why are rare recessives not rapidly eliminated?

8. Can any selective procedure ever completely eliminate deleterious mutations from a population?

9. In the case of a diploid organism, three allelic combinations must be considered in evaluating the selection coefficient. It has been said that the selection coefficient is based on the relative fitnesses and that any genotype can be chosen as the standard. However, a convention is used in choosing the standard. What is this convention?

10. What is the fitness of a recessive mutation that causes death of an organism before it can reproduce?

11. What is the selection coefficient for a recessive lethal?

12. What mutation rate can maintain an allele frequency of 0.0001 for a completely recessive allele?

13. What are the three possible effects of a mutation in a

heterozygote?

14. Are changes in allele frequencies from one generation to the next greater for smaller or larger populations?

15. In the absence of any counteracting forces random genetic drift causes an allele to become either fixed or lost. In terms of allele frequency, what does it mean to be fixed or lost?

16. What forces impede the fixation or loss of alleles that occurs in random genetic drift?

ANSWERS TO DRILL QUESTIONS

1. Mutation, migration, natural selection, random genetic drift.

2. Random genetic drift.

3. A population is in equilibrium when the rates of forward and back mutation for a particular gene are equal.

4. The allele frequency is the reverse mutation rate divided by the sum of the forward and reverse mutation rates.

5. Change in allele frequency per generation.

6. Yes; no.

7. If the recessive is rare, most recessive alleles are in heterozygotes and hence are not subject to selection.

8. No, because new mutations are continually arising.

9. The standard is generally the genotype with the maximum fitness in order that all values of relative fitness be less than or equal to one.

10. Zero, because the fitness is a measure of the ability of an organism to contribute genes to the gene pool.

11. Since the fitness is zero (see Question 9) and the selection coefficient is the difference between the fitnesses of the standard (defined as 1) and the mutant, the selection coefficient must be 1.

12. Use Equation 15-12. The mutation rate for a complete recessive is the square of the allele frequency, or 10^{-8}.

13. No effect, deleterious (partial dominance), and advantageous (heterozygote superiority).

14. Smaller. This is the phenomenon of genetic drift.

15. If lost, its allele frequency is zero; that is, the allele is no longer present in the population. If fixed, its allele frequency is 1; that is, it is the only allele at that locus present in the population--all members of the population are homozygous for the allele.

16. Mutation, migration, and natural selection.

ADDITIONAL PROBLEMS

1. Consider an organism that has, on the average, 3.3 children. A rare partially dominant mutation causes premature death during the child-bearing years, so that the average number of children is 1.1. What is the relative fitness of the heterozygote compared to the homozygous dominant?

2. A partially dominant mutation in the mouse causes the litter size of a heterozygote to be reduced, on the average to one third the normal value. If the allele frequency for the mutation is 3.3×10^{-5}, what is the mutation rate, assuming that the population is in equilibrium?

3. Mosquitoes are found worldwide and in fact the same species are found throughout the northern hemisphere. The genetics of mosquitoes has not been analyzed in any detail, but would you expect the allele frequencies of mosquitoes in Norway to be the same as in the United States, and would you expect mosquitoes in eastern United States to have the same allele frequencies as those in western United States?

4. Do you think that random genetic drift plays a role in determining allele frequencies in the two populations of houseflies east and west of the Rocky Mountains in the United States?

5. If the initial frequency of a mutant allele in a small population is 0.35, what is the probability that the population will be fixed for the wildtype?

6. The height of adult giraffes in eight different valleys in Kenya is being studied. Height is known to be a polygenic trait. It is observed that the mean values of the heights for the eight islands are: 72, 76, 79, 82 84, 91, 97, 103 inches, with a variation of rarely more than 2 inches. Give two explanations for these differences.

7. In a natural population of *Drosophila* the genotypes containing a particular allele *a* are found to occur with the frequencies: *AA,* 0.84; *Aa,* 0.14; *aa,* 0.02.
A and *a* form a partially dominant pair, so the three genotypes are identifiable by inspection. You have reasons to believe that these differences are not a reflection of random genetic drift but result from the fact that the three genotypes either have different abilities to survive to reproductive age or different brood sizes. One way to test this would be to study the lifetime and brood size of the

three genotypes separately. Give two reasons why this would not be a good way to answer the question, and also suggest an experimental approach that would be valid.

ANSWERS TO ADDITIONAL PROBLEMS

1. The fitness is defined in terms of ability to contribute alleles to the gene pool. Thus, in this case, the fitnesses are just the average number of offspring, so the relative fitness is simply the ratio of these numbers, or 1.1/3.3 = 0.33.

2. The relative fitness of the allele is 0.33, so that the selection coefficient against the heterozygote is 1 - 0.33 = 0.66. From Equation 15-13 the mutation rate is $(3.3 \times 10^{-5})(0.66) = 2.2 \times 10^{-5}$.

3. One would certain expect differences in allele frequencies for a variety of genes between Norway and the United States, because the populations have certain been isolated for innumerable generations and live in different environments. Mosquitoes in eastern and western United States are also geographically isolated by the Rocky Mountains, which divide the United States; thus, one would expect differences in the frequencies of any allele responsive to the environment.

4. If the eastern and western populations are considered to be single populations, random genetic drift will play little or no role in determining their allelic frequencies, because the populations are so large. However, each of these populations can certainly be divided into subpopulations (Manhattan houseflies versus Long Island houseflies), which certainly should show differences owing to random genetic drift. The effects might not be extremely large though, since insects are quite mobile, and migration would be an important countereffect.

5. Assuming that only two alleles are possible, mutant and wildtype, the initial frequency for the wildtype allele is 1 - 0.35 = 0.65. Thus, the probability of fixation of the wildtype is 0.65. See Problem 21 of the text.

6. (1) Different heights have different selective values in the different valleys. For example, the height of the acacia trees that giraffes feed on might select for extra neck or leg length, if the trees are especially high; or shortness might be associated with stronger legs that would enable them to kick some predator harder. (2) Random genetic drift acting on isolated populations that have not interbred for a very long time.

7. First, owing to statistical variation among individuals a very large number of flies would have to be studied; second and more significant, the three genotypes would not be competing with one another; third, the environments of each fly in the experiment might not be the same. The simplest

experiment would be to form a population consisting of equal numbers of each genotype, allow them to mate and reproduce, and look at the genotypic distribution of the progeny. This is straightforward because of the partial dominance of the alleles.

SOLUTIONS TO PROBLEMS IN TEXT

1. (a) The relative fitness is zero, because the genotype cannot contribute alleles to subsequent generations. (b) 1 − 0 = 1.0. (c) Fertility.

2. When an allele is rare, most alleles will occur in heterozygotes; and if the allele is recessive, by definition, the heterozygotes will not be affected by any selective force.

3. (a) Additivity means that the fitness of the heterozygote is the average of the fitnesses of the two homozygotes, or [1 (the standard) + 0.86]/2 = 0.93. (b) For Aa, 1 − 0.93 = 0.07; for aa, 1 − 0.86 = 0.14.

4. The equilibrium frequency of an X-linked allele is smaller than that of an autosomal recessive, because hemizygous males are subject to selection.

5. Decrease it, because more recessive homozygotes are formed, and these are subject to selection.

6. Random mating refers to random union of gametes. In a finite population, gametes uniting at random will sometimes join with those from relatives. In a population of size 1, random union of gametes must result in self-fertilization.

7. Use Equation 15-2, with $q_0 = 0$. Therefore, for strain A: 10^{-6}, 10^{-5}, and 10^{-4} for 10, 100, and 1000 generations, respectively. For strain B the mutation rate is 10-fold greater, so each of the values just given will also be 10-fold greater than in strain A.

8. Use Equation 15-3. (a) The equilibrium frequency of A is 4×10^{-8} divided by the sum $(4 \times 10^{-8} + 2 \times 10^{-6})$, or 0.0196. For a, the numerator of the expression is instead 2×10^{-6} and the frequency is 0.9804. (b) If the mutation rates were both increased by the same factor, both numerator and denominator would be multiplied by the same factor, so the equilibrium frequency would be the same.

9. Use Equation 15-4. The migration rate is $80/1000 = 0.08$ and there is one-way migration. $P = 1$, since the source aquarium contains albinos only, and $p_0 = 0$, since the recipient population had no albinos. Thus, after 10 generations the allele frequency would be 0.566, and after 50 generations it would be 0.985.

10. Use Equation 15-4, with $n = 100$, $p_n = 100$,

$p_0 = 0$, and $P = 1$. The result is $m = 0.0016$ per generation.

11. The fitness of A would have to be $(1.00 \times 0.96)^{1/2} = 0.9798$. Note that this value is roughly equal to the 0.98 expected with additive alleles.

12. (a) The different probabilities of survival are assumed to be the only differences in the ability of the different genotypes to reproduce. Thus, the relative fitnesses with A_3A_3 as the standard are $A_1A_1 = 0.60$, $A_1A_2 = 0.87$, $A_2A_2 = 0.87$, $A_2A_3 = 0.87$, $A_3A_3 = 1.00$, and $A_1A_3 = 0.80$. (b) Since the fitnesses of A_2A_2 and A_1A_2 are the same, A_2 must be dominant to A_1. Since the fitness of the A_2A_3 heterozygote is the same as that of A_2A_2, A_2 is dominant to A_3. A_1 and A_3 are additive, since the fitness of the heterozygote is the average of the fitnesses of the homozygotes. (c) Since A_3 is not dominant to either of the others, it will ultimately be fixed in heterozygotes.

13. (a) Selection coefficient is $1/0.98 - 1 = 0.02$. (b) Use Equation 15-5 with $w = 0.98$ and $p_0/q_0 = 60/40$. The values of p_n/q_n for $n = 10$, 50, and 90 generations are 1.2254, 0.5461, and 0.2434, respectively, which is the ratio of A to B. The fraction that is A is simply $A/(A + B) = A/[(A + A/(A/B)]$, or 0.5496, 0.3532, and 0.1958, respectively.

14. Use Equation 15-5 with $n = 40$, $p_0/q_0 = 1$ (equal amounts inoculated), and $p_{40}/q_{40} = 0.35/0.65$. The value of w is 0.9846. The selection coefficient is $1 - 0.9846 = 0.0154$.

15. Equation 15-5 can be used because the fitness of the heterozygote (0.5) is the square root of the product of the fitnesses of the homozygotes (square root of 1×0.25 is 0.5). The value of p_n/q_n is 0.3/0.7 and that for the initial state is 0.7/0.3. For $w = 0.5$ $n = 2.44$ generations.

16. Use Equation 15-15. For CH, the fitness is 0.46, and for ST the fitness is 0.89. Thus, for CH, $s = 0.54$ and $t = 0.11$ and the equilibrium frequency is $0.11/(0.54 + 0.11) = 0.17$. For ST, $s = 0.11$ and $t = 0.54$ and the equilibrium frequency is $0.54/0.65 = 0.83$.

17. (a) For a recessive lethal $s = 1$, so Equation 15-12 yields an equilibrium frequency of 0.002. (b) For the homozygote, the value is the square of that in part (a), or 4×10^{-6}.

18. For a recessive lethal the selection coefficient is 1. Thus, from Equation 15-12 the mutation rate = q^2 =

$1/118,000 = 8.47 \times 10^{-6}$.

19. (a) Use Equation 15-12 with s = 0.5. Thus, $(5 \times 10^{-5}/0.50)^{1/2} = 0.01$. (b) $5 \times 10^{-5}/0.02 = 0.0025$.

20. Use Equation 15-13 with s = 0.19. The mutation rate is $5 \times 10^{-5} \times 0.19 = 9.5 \times 10^{-6}$.

21. (a) 0.20, because as stated on page 555, the probability of ultimate fixation of a particular allele equals the frequency of that allele in the original population. (b) The effective population size.

22. Use Equation 15-15 with $N = 25$ and $n = 8$. $G = 0.15$.

23. Substitution into the equation given yields the following: numerator, $pq(1) + q^2(0) = pq$; denominator, $p^2(1) + 2pq(1) + q^2(0) = p^2 + 2pq = p(p + q + q) = p(1 + q)$, so the ratio is $q/(1 + q)$. Note that $1/q' = 1 + 1/q$, so $1/q_1 = 1 + 1/q_0$; $1/q_2 = 1 + 1/q_1 = 2 + 1/q_0$; ...; $1/q_n = n + 1/q_0$, which when rearranged gives the equation stated.

CHAPTER 16

Quantitative Genetics

CHAPTER SUMMARY

Many traits that are important in agriculture and human genetics are determined by the effects of several or many genes acting in combination with influences of the environment. Traits with such complex determination are multifactorial traits, and their study is known as quantitative genetics. Three types of multifactorial traits may be distinguished--quantitative, meristic, and threshold traits. Quantitative traits are those that are expressed according to a continuous scale of measurement, like height or weight. Meristic traits are traits that are expressed in whole numbers, like the number of grains on an ear of corn. Threshold traits have an underlying multifactorial risk or liability and are either expressed or not expressed in each individual; an example is diabetes. The concepts used in understanding multifactorial traits are the same for all three types.

The relation between the phenotypes that occur in a population and the frequency of the phenotypes is called the distribution of the trait. Many quantitative and meristic traits have a distribution that approximates the bell-shaped curve of a normal distribution. A normal distribution can be completely described by two quantities--the mean and the variance. The standard deviation of a distribution is the square root of the variance. In a normal distribution, approximately 68 percent of the individuals will have a phenotype within one standard deviation from the mean, and approximately 95 percent of the individuals will have a phenotype within two standard deviations from the mean. However, the mere observation that a trait is normally distributed conveys no information about the degree or nature of genetic determination of the trait.

Variation in phenotype of multifactorial traits among individuals in a population derives from four principal sources: (1) variation in genotype, which is measured by the genotypic variance, (2) variation in environment, which is measured by the environmental variance, (3) variation resulting from the interaction between genotype and environment (G-E interaction), and (4) variation resulting from nonrandom association of genotypes and environments (G-E association). If G-E interaction and G-E association can be neglected, then the total variance in phenotype in a population is expected to be the sum of the genotypic variance and the environmental variance. These components of variation can be estimated from the variance in phenotype observed among inbred lines and between the F_1 and backcrosses.

The ratio of genotypic variance to the total phenotypic variance of a trait is called the broad-sense heritability of the trait, and this quantity is useful in predicting the outcome of artificial selection practiced among clones, inbred lines, or varieties. When artificial selection is practiced in a randomly mating phenotype, then a second type of heritability, the narrow-sense heritability, is used for prediction. The value of narrow-sense heritability of a trait can be estimated from the resemblance in phenotype among groups of relatives.

One common type of artificial selection is called truncation selection, in which only those individuals that have a phenotype above a certain value (the truncation point) are saved for breeding the next generation. Artificial selection usually results in improvement of the selected population; indeed, after many generations the mean of the selected population may be many standard deviations greater than the mean of the original population. However, for two reasons progress often slows or ceases when selection is carried out for many generations: (1) some of the favorable genes become nearly fixed in the population, which decreases the narrow-sense heritability, and (2) natural selection may counteract the artifical selection.

The principles of quantitative genetics can also be applied to the analysis of behavioral traits in humans and other animals. Hoqever, because of the possibility of G-E interaction and G-E association, such applications require great care in the execution and great caution in the interpretation.

BOLD TERMS

additive variance, artificial selection, broad-sense heritability, continuous traits, correlation coefficient, correlated response, covariance, distribution, dominance variance, effective number of loci, environmental variance, fraternal twins, G-E association, G-E interaction, genotype-environment association, genotype-environment interaction, genotypic variance, heterosis, inbreeding depression, identical twins, individual selection, liability, mean, meristic traits, narrow-sense heritability, normal distribution, path diagram, path coefficients, paths, prediction equation, quantitative genetics, quantitative traits, selection limit, standard deviation, threshold, threshold traits, total variance, truncation point, variance.

ADDITIONAL INFORMATION

The mechanics of analyzing multifactorial traits is quite different from the analysis of single-gene traits seen in early chapters of the text. However, one must not be misled into thinking that the genes affecting quantitative traits are any different from those affecting simple traits. There is nothing unusual about these genes, and they follow Mendelian patterns of inheritance, with multiple alleles, dominance relations, and so forth. The real point is that when several genes affect a trait and when one does not know what these genes are, the procedures of Mendelian analysis simply cannot be carried out. A further complication is the profound effect of environment on gene expression. Again, there is nothing special about these genes and, in this respect, environment often affects the phenotype of a single-gene trait. However, with a single-gene the effect of environment on expression of that gene can be studied in isolation, whereas with multifactorial traits the effect of environment can be different for each gene. Furthermore, the genes contributing to quantitative traits do not always contribute equally; in fact, it has become clear that in determining the phenotype of most traits a few genes have large effects (major genes) and many genes have small effects (minor genes). The result of this complexity is that the analysis of multifactorial traits relies heavily on statistics, and hence statistical manipulations account for a significant fraction of this chapter.

The study of quantitative genetics is important as a means of both understanding the nature of the inheritance of multifactorial traits and of manipulating populations to achieve desired results (that is, breeding). In breeding, it is important to be able to separate environmental and genetic effects, since one can only select for genotypes. This is why one evaluates quantities such as the narrow-sense heritability, which provides a means of predicting how successful a particular selective procedure will be. It is worth noting that one way to evaluate the narrow-sense heritability is to observe changes in mean values following particular selective procedures. For example, Equation 10, which relates the mean of a population, the mean of a group selected as parents of a new population, and the mean of the offspring, can be used to determine the narrow-sense heritability by measuring M, M', and M^*, as well as to predict the results of a breeding program, once h^2 is known. The major method for evaluating the narrow-sense heritability is by studying traits in relatives, using the relations shown in Table 16-6.

At one time it was believe that the most precise values of the heritability of certain traits could be obtained by comparing monozygotic and fraternal, since they have the identical genotype and their variance is all environmental. The idea is to understand the effect of environment on certain traits. The method was to compare the variances of monozygotic and fraternal twins. To avoid the cumbersome notation with squares for the variance, let us, for the moment, denote the variance of monozygotic and fraternal twins by v_m and v_f, respectively, and use v_e and v_g for the environmental and genetic variance. For monozygotic twins, the variance is all environmental, or $v_m = v_e$. Fraternal twins have half of their genes in common, so their genetic variance is half that of unrelated individuals, that is, $v_f = (1/2)v_g + v_e$. The difference between the phenotypic variances of monozygotic and fraternal twins is $v_f - v_m = (1/2)v_g + v_e - v_e = (1/2)v_g$. Since $H^2 = v_g/v_{total}$, $H^2 = 2(v_f - v_m)/v_{total}$. However, the use of twins is not as straightforward as has been thought. An implicit assumption is the the environmental variance is the same for monozygotic twins and for fraternal twins of the same sex, because the two individuals are born at the same time, live together, and totally share their early lives. However, it seems likely that monozygotic twins are treated more similarly than fraternal twins, because they react more similarly. Thus, v_e of the fraternal twins may be larger than the value for the monozygotic twins, so the difference $v_f - v_m$ is greater and the formula just given overestimates heritability. This has led to certain incorrect notions about the degree to which certain traits have a genetic basis. A nearly ideal case is to compare twins reared in the same family with twins raised in separate families, but the number of cases of that type studied is quite limited. The message is that the results of quantitative analyses based on twin studies should be accepted cautiously.

DRILL QUESTIONS

1. What is meant by a quantitative trait?
2. Does a quantitative trait differ from a multifactorial trait?
3. If a quantitative trait is determined genetically by the alleles of a single gene, what other factor(s) might influence the trait?
4. What kind of quantitative trait is the weight of wool produced by sheep?
5. If two animals have 38.2-cm tails, one has a 32.4-cm tail, and a fourth has a tail 36.8 cm long, what is the mean tail length? Is this value representative of the tail length of a larger population of the animal?
6. What is the variance of the sample in Question 5?
7. How are the standard deviation and the variance related?
8. What is the standard deviation of tail length of the animals in Question 5?
9. Is there any circumstance in which the variance of a distribution will be zero?
10. What is the value on the x axis corresponding to the peak of a normal distribution called?
11. Two normal distributions are being compared. They have the same mean but different widths. Which has the larger variance, the wider or the narrower distribution?
12. If you can calculate a mean and variance for a distribution, is the distribution necessarily a normal distribution?
13. The lengths of adult earthworms in a given field is being studied. The mean length is 6.1 cm and the variance is 0.04 cm. If 1000 worms are studied, roughly how many will have a length in the interval 5.9 to 6.3 cm?

14. Several thousand seeds are collected from fruit obtained from apparently identical plants in the Brazilian jungle. You plant them in your greenhouse and discover that some plants produce considerable heavier fruit than others. What type of variance would be obtained from a measurement of fruit weight?

15. Very expensive hybrid corn seeds are purchased from a reputable seed supplier for use on a farm cooperative. The seeds are planted on each farm, but with great disappointment it is found that the yield per acre in some farms is quite low and in others quite large. The farmers calculate the variance and send this value to the seed supplier, complaining that the seeds were defective because the variance was five times as large as was advertised. What will the supplier tell the farmers about their numbers?

16. In comparing a measure of a quantitative trait in the F_1 obtained from two highly inbred strains and in the F_2, which set of progeny provides the value of the environmental variance alone?

17. How is the effective number of loci related to the actual number of loci that determine a particular trait?

18. If you knew the genetic variance for a heterozygote, what two measurements would have to be made in order to calculate the effective number of loci?

19. What is the broad-sense heritability of a population of individuals homozygous for all genes determining a trait?

20. If one artificially selects for a particular trait and if the selection is successful, will the broad-sense heritability of the trait increase or decrease with the number of generations since the beginning of the selection? How will the narrow-sense heritability vary?

21. What is the essential difference between the variance and the covariance? Do not give a mathematical definition.

22. What does the narrow-sense heritability tell about the inheritance of a trait?

23. If the covariance for a trait compared between first cousins is 1.23, what is the value of the additive variance? What is the narrow-sense heritability if the total phenotypic variance is 12.43?

ANSWERS TO DRILL QUESTIONS

1. Any trait that is not determined only by alternate genotypes at a single locus.

2. No. The two terms are used interchangeably.

CHAPTER 16   227

3. Environmental factors.

4. A continuous trait.

5. (1/4) (38.2 + 38.2 + 32.4 + 36.8) = 36.4 cm. No, but if that is the only data available, it can be used.

6. Subtract the mean from each value and square the difference. Sum these differences (= 22.64) and divide by the number of samples (4) minus 1. Thus, the variance is 22.64/3 = 7.55.

7. The standard deviation is the square root of the variance.

8. The variance was 7.55, so the standard deviation is the square root of 7.55, or 2.75.

9. Only if all values were the same. In this case, each value would be the mean, and the differences between individual values and the mean would each be zero.

10. The mean.

11. The wider one.

12. No. You can calculate a mean and a variance for any set of numbers. However, the distribution of values must have certain features to be called a normal one.

13. The variance is 0.04, so the standard deviation is 0.2. If the lengths actually conform to a normal distribution, 68 percent (680 worms) should have a length between 5.9 and 6.3 cm.

14. Presumably the environment in a greenhouse is constant, so the variance would be genetic. This would be expected since no information is available about whether the original seeds came from either true-breeding or genetically identical plants.

15. If the advertiser is a knowledgable geneticist, the farmers will be told that they are observing different genotype-environment interactions in the different fields.

16. The genotypic variance of an F_1 is zero, because all individuals have the same genotype. Thus, the variance of the F_1 is solely the environmental variance.

17. It is less than or equal to the actual value.

18. Measurements of the mean values of the trait for the two homozygous parental types, because D in Equation 16-7 is the difference between these values.

19. Zero, because the genetic variance, which appears in the numerator of the expression for the broad-sense heritability, is zero.

20. The broad-sense heritability will decrease, because as a favorable allele is either fixed or lost, the genotype variance decreases. The narrow-sense heritability is always less than the broad-sense heritability, so it will also decrease.

21. The variance measures the degree of variation within a collection of data. The covariance applies to two sets of data and is a measure of the correlation between pairs of values in the different sets.

22. Nothing, but it does tell how the trait can be changed by selection.

23. From Table 16-6, the additive variance $A = 8(1.23) = 9.84$, and the narrow-sense heritability is $9.84/12.43 = 0.79$.

ADDITIONAL PROBLEMS

1. The variance of egg number produced in one year by the F_1 obtained from a mating between two highly inbred strains of canaries is 1.24 and that of the F_2 is 2.74. What is the genetic variance for the F_2? What information do you have about the number of genes that determines egg number?

2. Two highly inbred strains of an animal are mated produce an F_1, and then the F_1 individuals are mated to produce an F_2. The variances for a particular quantitative trait are determined, and it is found that the variance for the F_1 is 3.4 and that of the F_2 is 1.2. What can you conclude from these values?

3. Except in the case of codominance or partial dominance, the phenotype of the heterozygote is that of the homozygous dominant, as seen in Chapter 1. In view of this phenomenon, what explanation can be given for heterosis?

4. It is a well-known phenomenon that monozygotic twins tend to get the same diseases caused by bacteria and viruses than do siblings that are not twins. Studies of a variety of such diseases indicate that if one twin gets the disease, the probability that the other will get it is between 80 and 90 percent, but rarely 100 percent. For other siblings the probability is less than 50 percent, except for the most contagious diseases. Suggest a reason for the high correlation, and also account for the fact that the probability is not 100 percent.

5. A particular malformation in humans is being studied in order to determine its mode of inheritance. The risk in sibs of affected individuals, 10 percent, is the same as that in offspring. The risk in nieces and nephews is 5 percent, and in first cousins, 2.5 percent. Do you think that the trait is multifactorial or an autosomal dominant with reduced penetrance? (You cannot be sure but the relative values of the risk might give you a hint.)

ANSWERS TO ADDITIONAL PROBLEMS

1. For the F_1 the genotypic variance is 0, so 1.24 represents the environmental variance. Thus, the genotypic variance of the F_2 is 2.74 - 1.24 = 1.50. A measurement of the values of the variance for these two sets of progeny, without knowing the values for the parents, does not tell you anything about the number of genes.

2. The variance of the F_1 is environmental variance only and that of the F_2 is genotypic + environmental variance. Therefore, if the environments of the F_1 and F_2 were the same, the total variance of the F_2 would have to be greater than or equal to that of the F_1. Since that is not the case, you can only conclude that the numbers cannot be compared, because different environments have been present.

3. Heterosis occurs only with multifactorial traits. Highly inbred strains are usually homozygous for many deleterious alleles. When crossing these inbred lines, the homozygosity of these alleles is eliminated by the formation of the heterozygous state. The favorable dominant then determines the trait.

4. There is no doubt that different people have different abilities to resist disease and that this is a reflection of the "power" of their immune systems. Monozygotic twins, having the same genotype, have a common potential for developing immunity and therefore for resisting infection. Thus, because both members will usually be exposed they have about the same probability of contracting disease. However, their lifelong environments have not been identical; for example, they probably have not always eaten the same things, nor have they always been exposed to the same microorganism, which may have antigens in common with other disease-causing organisms. Thus, their total ability to respond to a new microorganism is not the same.

5. Note that the risk drops by a factor of two with each step of a more remote relationship. Such a fixed factor would be very unlikely with a multifactorial trait, suggesting that the trait is an autosomal dominant with reduced penetrance. Studies with twins would give the most information. If it were an autosomal dominant, the risk in monozygotic twins should be about twice that of fraternal twins and probably much less than 50 percent; also, if it were multifactorial, the relative risk with monozygotic twins would probably be very high.

SOLUTIONS TO PROBLEMS IN TEXT

1. Each variety is stated to have different productivity in two different regions (environments), so this is an example of genotype-environment interaction.

2. The gametes forming child 1 could have come from his parents, the other parents, or either of the other two

male-female combinations. The same is true of child 2. Therefore, the children are full siblings.

3. The number of mice is 10, the total number of mice is 104, so the mean = 104/10 = 10.4. The variance = 24.4/(10 - 1) = 2.71. The standard deviation = $(2.71)^{1/2}$ = 1.65.

4. (a) To determine the mean, multiply each value for the pounds of milk produced (using the midpoint value) by the number of cows and divide by 304 (the total number of cows) to obtain a value of the mean of 4032.9. The variance is 693,633.8, and the standard deviation, which is the square root of the variance, is 832.8. (b) The range that includes 68 percent of the cows is between the mean minus the standard deviation and the mean plus the standard deviation, that is, 4033 - 833 = 3200 to 4033 + 833 = 4866. (c) Rounding to the nearest 500 yields 3000 to 5000. The observed number in this range is 239; the expected number is (0.68)(304) = 207. (d) For 95 percent of the animals the range is within two standard deviations or 2367 to 5698. Rounding off to nearest 500 yields 2500 to 6000. The observed number is 296, which compares to the expected number of (0.95)(304) = 289.

5. (1) 15 to 21 is the range within one standard deviation of the mean, or 68 percent. (2) 12 to 24 is within two standard deviations of the mean, or 95 percent. (3) Since a normal distribution is symmetric about the mean, the number lower than one standard deviation below the mean is (1 - 0.68)/2 = 0.16. (4) More than 24 leaves is greater than two standard deviations above the mean, or (1 - 0.95)/2 = 0.025. (5) Between 21 and 24 leaves is between one and two standard deviations above the means, or 0.16 - 0.025 = 0.135, because 0.16 have more than 21 leaves and 0.25 have more than 24 leaves.

6. (a) This problem is made simple by the fact that dominant and recessive alleles are equally frequent. The probability of any combination of alleles is $(1/2)^6$ = 1/64. There is only one combination with a phenotype of 0, namely, the complete recessive, so the frequency is 1/64. For a phenotype of 1, for each gene, there are two possible genotypes, namely, 1 homozygous dominant and two heterozygous. There are three genes, so there are nine combinations, and the frequency for the phenotype of 1 is 9/64. For a phenotype of 2, there is a 3/4 probability of being homozygous dominant or heterozygous for one gene, and (3/4)(3/4) = 9/16 for two genes. For only two genes, we must multiply by 1/4, the probability of being homozygous recessive for the third gene. Therefore, the frequency is 27/64 for a phenotype of 2. The frequency of a phenotype of 3 is the remaining frequency or 1 - (1/64 + 9/64 + 27/64) = 27/64. (b) If each dominant allele has a frequency of 1/4, the recessive has a frequency of 3/4. Therefore, the frequency for a phenotype of 0 is $(3/4)^6$ = 729/4096. For a phenotype of 1, we note that the frequency of a heterozygote or a homozygous dominant is 2(3/4)(1/4) + (1/4)(1/4) = 7/16,

which must be multiplied by $(3/4)^4$ for the other four recessive genes, and by three since there are three genes, yielding a frequency of 1701/4096. For a phenotype of 2, $(7/16)(7/16)(3/4)^2(3)$ = 1323/4096. Subtraction of the sum of the frequencies from 1 yields 343/4096 for a phenotype of 3. (c) Part (a): mean, 2.25; variance, 0.56; standard deviation, 0.75. Part (b): mean, 1.3125; variance, 0.738; standard deviation, 0.86.

7. (a) The genotypic variance is 0 because all F_1 individuals have the same genotype. (b) With additive alleles, the mean of the F_1 individuals should equal the average of the parental strains.

8. For the F_1 the total variance equals the environmental variance, which we are told is 1.46. For the F_2 the total variance, which is 5.97, is the sum of the environmental variance (1.46) and the genotypic variance, so the genotypic variance is 5.97 - 1.46 = 4.51. The broad-sense heritability is the genotypic variance (4.51) divided by the total variance (5.97), or 4.51/5.97 = 0.76.

9. Using Equation 16-7, the minimum number of genes affecting the trait is D^2/8(environmental variance) = 44.

10. Using Equation 16-10, h^2 = (2.33 - 2.26)/(2.37 - 2.26) = 0.64.

11. (a) See the lower display equation on page 586. 12 + 0.20(15 - 12) = 12.6 grams (b) 12 + 0.20(9 - 12) = 11.4 grams.

12. The mean of the 20 numbers given is 81, and standard deviation is 13.8. The mean of the upper half of the distribution could be calculated directly (by summing the numbers above the mean and averaging them), or the expression given could be used. Using that expression, the mean of the upper half of population is 81 + 0.8(13.8) = 92. Using the expression at the bottom of page 586, the expected weaning weight of the offspring is 81 + 0.2(92 - 81) = 83.2.

13. Use Equation 16-10. M = 700. Since there is a 50-gram increase per generation for five generations, the total increase is 250 grams, which is $M*$ - M. Therefore, 0.8 = (M' - M)/250 = 0.8 and M', the expected mean weight-gain after selection is 900 grams.

14. Use Equation 16-10. The value of the narrow-sense heritability is needed; this can be calculated as h^2 = 10,000/40,000 = 0.25. The standard deviation is 200 micrograms, so the selection is for individuals with a mean weight of 2000 + 2(200) = 2400. Thus, the expected mean = 2000 + 0.25(2400 - 2000) = 2100 micrograms.

15. Use Equation 16-10. M, the number of trials for the original population, is given as 10.8. The parents used were those that could learn in 5.8 trials, which is $M*$. The resulting average number of trials, which is M', was 8.8.

Therefore, the narrow-sense heritability $h^2 = (8.8 - 10.8)/(5.8 - 10.8) = 0.40$.

16. (a) Use Equation 16-10 with $M = 10.8$, $M^* = 5.8$, and $M' = 9.9$. In this case, $h^2 = (9.9 - 10.8)/(5.8 - 10.8) = 0.18$. (b) The result is consistent with that on Problem 15 because of the differences in the variance. In Problem 15, the additive variance was $0.4 \times 4 = 1.6$. In the present case the phenotypic variance is 9. If the additive variance did not change, then the expected heritability would be $1.6/9 = 0.18$, which is observed. However the problem has been idealized somewhat in that in real examples the additive variance would also be expected to change with a change in environment.

17. An incidence of 1 in 10,000 is an incidence of 0.01 percent. Read upward from 0.01 to the line for $h^2 = 0.80$. The value on the y axis is 1 percent.

18. An incidence of 1 in 2000 is 0.05 percent. Read upward from an x value of 0.05 to the y value of 1. The intersection is on the curve for $h^2 = 0.60$.

19. The risk increases as the incidence of the trait in the population increases, because a higher incidence implies a lower threshold, and a lower threshold increases the risk for everyone, including the first-degree relatives of affected individuals.

20. $r = 0.57$, in actual data; $r = 0.60$ for mismatched data. Note that the standard deviations calculated from the first litter (1.65) and from the second litter (2.11) are independent of how the numbers are paired. The correlation coefficients differ (though not by much in this case) when the numbers are rearranged, because different pairs of numbers are multiplied. Note also that when calculating a correlation coefficient, it is essential to keep track of the signs of the products; the difference between an individual value and the mean can be either positive or negative. Thus, for the first litter, for which the mean is 10.4, the differences are, in order: 0.6, -1.4, 2.6, -0.4, -1.4, -2.4, -0.4, 0.6, -0.4, and 2.6. For the second litter, for which the mean is 10.3, the differences are 1.7, -0.3, 1.7, 1.7, -0.3, -2.3, -4.3, 1.7, -1.3, and 1.7. The products are: -0.18, -2.38, 4.42, 0.12, 3.22, 10.32, -0.68, -0.78, -0.68, and 4.42. Thus, the correlation coefficient is $1/(10 - 1) = 1/9$ times the sum of the products. The sum, of course, differs for the two orders of numbers.

21. Let P represent parent-offspring covariance and S represent the full sib covariance. From Table 16-6, $D = 4(S - P)$ and $A = 2P$. H^2 is estimated as $(A + D)/$(total phenotypic variance), and $h^2 = A/$(total phenotypic variance).

CHAPTER 16

22. From Table 16-6 the covariance is $A + D$, so from the definition of broad sense-heritability, H^2 = covariance/variance. Thus, for tooth number $H^2 = 1.76/1.84 = 0.96$, and for the length of the longest spine, $H^2 = 0.91/3.17 = 0.29$.

23. The offspring are genetically related as half-siblings, so using Table 16-6, the theoretical covariance is $A/4$.

24. The mother-offspring covariance would be greater than the father-offspring covariance.

25. (a) If there is no selection of any kind, the mean of the population does not change from one generation to the next in a large population. Thus, the mean is 100. (b) Among sperm-bank pregnancies, average of males and females is $(130 + 100)/2 = 115$. Using the lower equation of page 586, the mean of offspring from such pregnancies is $100 + 0.3(115 - 100) = 104.5$. Among non-sperm-bank pregnancies, mean of offspring is 100, as stated in (a). The next generation consists of 10 percent sperm-bank offspring and 90 percent non-sperm-bank offspring, so the expected mean of all progeny is $0.9 \times 100 + 0.1 \times 104.5 = 100.45$. Since this change requires one generation, or 25 years, the average increase will be $(100.45 - 100)/25 = 0.018$ IQ points per year. This calculation indicates that such a recent proposal of the use of sperm banks from Nobel laureates and other people of superior ability would have such a small effect that the proposal is absurd.

26. (a) The narrow-sense heritability = (additive variance)/(total variance) = $2pq^3/p^2(1 + q)$ = $2q/(1 + q)$. The broad-sense heritability is (additive variance + dominance variance)/total variance = $(2pq^3 + p^2q^2)/pq^2(1 + q) = (2q + p)/(1 + q)$. Since $p + q = 1$, this expression reduces to 1. (b) For the values of q given, the values of the narrow-sense heritability are 1.0, 0.67, 0.18, 0.10, 0.02, 0.01, and 0.002, respectively. (c) When the recessive allele is rare, most homozygotes will come from matings between heterozygotes. Hence there will be no parent-offspring resemblance with respect to the trait.